Student's Choice

Regents Review

Geometry

Henry Gu

Mathematics Teacher
John Dewey High School
Brooklyn, New York

Disclaimer: The contents of this book are the author's alone and not those of the New York City Department of Education.

Author: Henry Gu
Editor: Christopher Gu

www.hsmathreview.com

ISBN-10: 1453709991   ISBN-13: 9781453709993

"Everything should be made as simple as possible, but not simpler."

- Albert Einstein

Preface

Nowadays, students are struggling to learn math and pass exams. They are overwhelmed with information from lengthy textbooks, review books, and many math websites. With limited time, students cannot benefit from all these resources.

Our students need only one concise book to help them review and prepare for the Geometry Regents exam. This is the book!

"No more. No less. Just right."

This book is structured in three parts:
1. A Geometry review that will help students remember all the key topics and build their problem solving skills through the use of examples.
2. A practice section with real Regents questions.
3. Answers and explanations.

The topics for the practice questions correspond to the sections in the Geometry review. Students can easily refer back to the matching review sections, while they are doing the practice.

This review book is geared towards helping students succeed with high scores on the Regents exams. I have already used these review sheets with my own Regents classes and I have seen firsthand that their performance is significantly higher than the statewide average. Both students and teachers like these review sheets because they are straightforward and practical.

For updates and information on additional titles, please visit our website www.hsmathreview.com.

Acknowledgement

Thanks to the teachers and students at John Dewey High School who have already used these review sheets for their own Regents review and have achieved excellent scores.

Thanks to my family for their unconditional love and support.

Dedication

This book is dedicated to all the students taking the Regents exams. I wish you the best of luck!

Contents

Contents

I. LOGIC

1. Negation: not, Symbol ~
e.g.   Statement p: I am a student.   T
      Negation ~ p: I am not a student.   F

Truth Values:

p	~ p
T	F
F	T

2. Conjunction: and, Symbol ^
The conjunction p ^ q is true only when both parts are true.

Truth Value:

p	q	p ^ q
T	T	T
T	F	F
F	T	F
F	F	F

3. Disjunction: or, Symbol v
The disjunction p v q is false only when both parts are false.

Truth Value:

p	q	p v q
T	T	T
T	F	T
F	T	T
F	F	F

e.g. p: 10 is divisible by 2.   T
    q: 10 is divisible by 3.   F
    p ^ q: 10 is divisible by 2 and 10 is divisible by 3.   F
    p v q: 10 is divisible by 2 or 10 is divisible by 3.   T
    p ^ ~q: 10 is divisible by 2 and 10 is not divisible by 3. T

4. Conditional Statements
(1) **Original** p ----> q   (If p then q.)
e.g. If it is snowing, then the school is closed.
(2) **Inverse** ~p ---> ~q (If not p then not q.)
e.g. If it is not snowing, then the school is not closed.
(3) **Converse** q ---> p (If q then p.)
e.g. If the school is closed, then it is snowing.
(4) **Contrapositive** ~q ---> ~p (If not q then not p.)
e.g. If the school is not closed, then it is not snowing.

Statements (1) and (4) are logically equivalent.
Statements (2) and (3) are logically equivalent.

5. Biconditional Statements
p <----> q (p and q have the same truth value.)
e.g. All the definitions are biconditional statements.
e.g. Two lines are perpendicular if and only if they
form right angles.

II. POSTULATES

1. Postulates of Equality
(1)   a = a   **Reflexive Property;**

e.g.

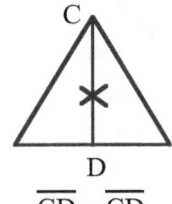

$$\overline{AB} \cong \overline{AB} \qquad \overline{CD} \cong \overline{CD}$$

(2)   If a = b, then b = a   **Symmetric Property;**
(3)   If a = b and b = c   **Transitive Property;**
    then a = c
(4)   If a = f(b) and b =c   **Substitution Postulate;**
    then a = f(c)
e.g.   If a = 2b and b = c
    then a = 2c ;
(5)   If a = b and c = d   **Addition Postulate;**
    then a + c = b + d
e.g.   If a = b then a + c = b + c ;
(6)   If a = b and c = d   **Multiplication Postulate;**
    then ac = bd

e.g. If a = b then $\dfrac{a}{2} = \dfrac{b}{2}$; Halves of equal quantities
             are equal (**Division Postulate**)

2. Postulates of Inequality
(1)   If a > b and b > c   **Transitive Property;**
    then a > c

(2)   If a > b and b = c   **Substitution Postulate;**
    then a > c

(3)   If a > b and c > d   **Addition Postulate;**
    then a + c > b + d
e.g.   If a > b then a + c > b + c ;

(4)   If a > b and c > 0   **Multiplication Postulate;**
    then ac > bc
e.g.   If a > b then 2a > 2b ;

3. Partition Postulate
    A whole is equal to the sum of all its parts.
e.g   AD = AB + BC + CD
    A whole is greater than any of its parts.
e.g. AD > AB , AD > BD , AD > BC

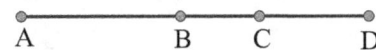

III. DEFINITIONS

1. Angles

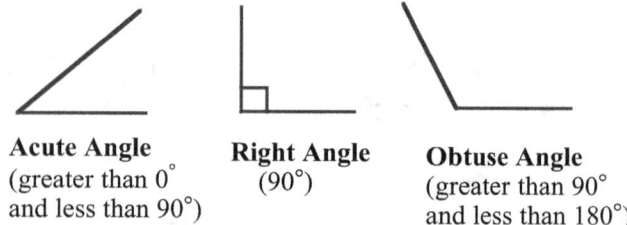

Acute Angle
(greater than $0°$
and less than $90°$)

Right Angle
$(90°)$

Obtuse Angle
(greater than $90°$
and less than $180°$)

If $\angle A$ and $\angle B$ are **complementary**, then
      $m\angle A + m\angle B = 90$ vice versa.
If $\angle A$ and $\angle B$ are **supplementary**, then
      $m\angle A + m\angle B = 180$ vice versa.
e.g. A linear pair of angles are supplementary.

2. A **midpoint** divides a line segment into two
congruent segments.

3. A **bisector of a segment** divides the segment into
two congruent segments.

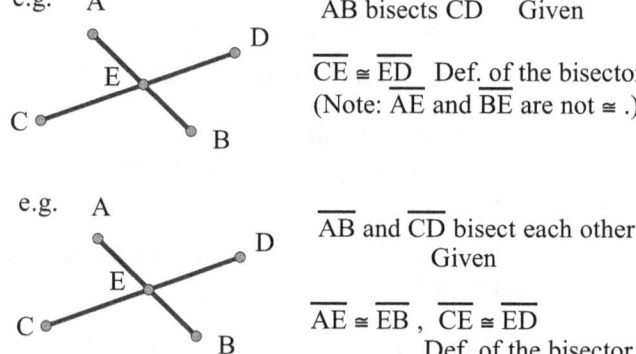

$\overline{AB}$ bisects $\overline{CD}$ Given

$\overline{CE} \cong \overline{ED}$ Def. of the bisector
(Note: $\overline{AE}$ and $\overline{BE}$ are not $\cong$.)

$\overline{AB}$ and $\overline{CD}$ bisect each other
                  Given

$\overline{AE} \cong \overline{EB}$, $\overline{CE} \cong \overline{ED}$
            Def. of the bisector

4. A **bisector of an angle** divides the angle into two
congruent angles.

5. **Perpendicular lines** intersect to form right angles.

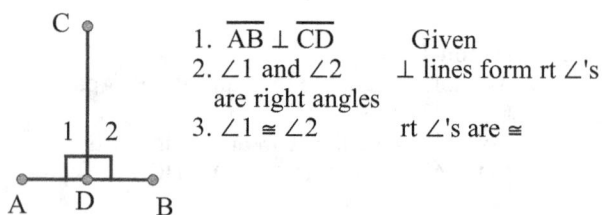

1. $\overline{AB} \perp \overline{CD}$ Given
2. $\angle 1$ and $\angle 2$ $\perp$ lines form rt $\angle$'s
   are right angles
3. $\angle 1 \cong \angle 2$ rt $\angle$'s are $\cong$

6. **Parallel lines** are in the same plane and do not
intersect.

IV. THEOREMS

1. Vertical angles are congruent.

e.g.

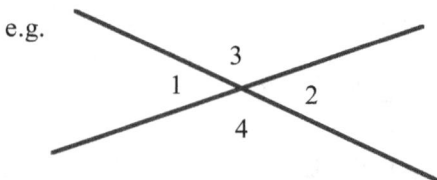

$\angle 1 \cong \angle 2$ and $\angle 3 \cong \angle 4$ (vertical angles are $\cong$.)
$m\angle 1 + m\angle 4 = 180$ and $m\angle 2 + m\angle 4 = 180$
(A linear pair of angles are supplementary.)

2. All right angles are congruent.

3. If two angles are congruent, then their complements
are congruent.

4. If two angles are congruent, then their supplements
are congruent.

5. Perpendicular Lines
(1). Perpendicular lines form right angles.
(2). If a point is on the perpendicular bisector of a line
segment, then it is equidistant from the endpoints of
the line segment, and vice versa.
(3). If two points are each equidistant from the
endpoints of a line segment, these points determine the
perpendicular bisector of the segment.

6. Parallel Lines
Parallel lines are everywhere equidistant.

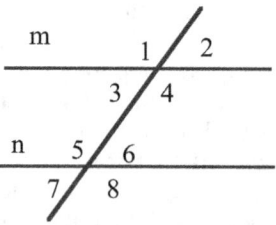

If line m ‖ line n, then alternate interior angles are $\cong$.
        $\angle 3 \cong \angle 6$ and $\angle 4 \cong \angle 5$
If line m ‖ line n, then corresponding angles are $\cong$.
    $\angle 1 \cong \angle 5$, $\angle 2 \cong \angle 6$, $\angle 3 \cong \angle 7$, $\angle 4 \cong \angle 8$
If line m ‖ line n, then interior angles on the same
side of the transversal are supplementary.
        $m\angle 3 + m\angle 5 = 180$
        $m\angle 4 + m\angle 6 = 180$

V. TRIANGLES

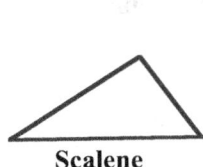  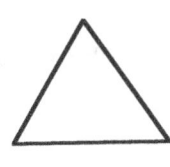

    Scalene   **Isosceles**   **Equilateral**

(no congruent sides) (2 congruent sides) (3 congruent sides)

1. The sum of the three interior angles is 180° ;
   The exterior angle is equal to the sum of 2 nonadjacent
   interior angles.

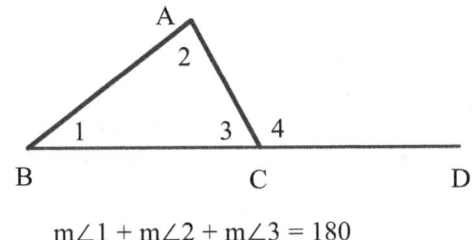

$$m\angle 1 + m\angle 2 + m\angle 3 = 180$$

$$m\angle 1 + m\angle 2 = m\angle 4$$

2. Triangle Inequalities

(1) In any triangle the greater side is opposite the greater angle, and vice versa.

(2) Any side is greater than the difference of the other 2 sides and less than the sum of them.
$$\left|s_1 - s_2\right| < s_3 < \left|s_1 + s_2\right|$$
e.g. If the two sides of a triangle are 3 and 5, then the 3rd side s_3 is $\left|3 - 5\right| < s_3 < 3 + 5$, which is $2 < s_3 < 8$

(3) Any exterior angle is greater than either nonadjacent interior angle.

3. Median, Altitude, and Angle Bisector

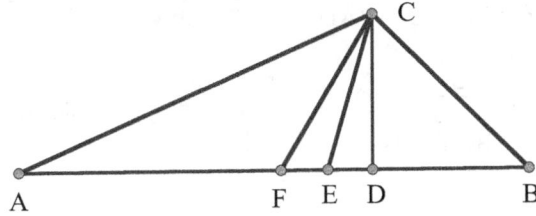

Given: $\overline{CD}$ is an altitude, $\overline{CE}$ is an angle bisector, and
     $\overline{CF}$ is a median.

$\overline{CD} \perp \overline{AB}$	Def. of altitude
$\angle ACE \cong \angle BCE$	Def. of angle bisector
F is the midpoint of $\overline{AB}$	Def. of median
$\overline{AF} \cong \overline{BF}$	Def of midpoint

Concurrence (Intersect in One Point)

(1) Centroid:
The medians of a triangle are concurrent.
(It is the Center of Gravity)
The centroid divides each median in the ratio 2 to 1.

(2) Incenter:
The angle bisectors of a triangle are concurrent.
(It is the Center of the Inscribed Circle
      --- equidistant from each side)

(3) Orthocenter:
The altitudes of a triangle are concurrent.
The orthocenter of an obtuse $\triangle$ is outside of the triangle.

(4) Circumcenter:
The perpendicular bisectors of the sides of a triangle are concurrent.
(It is the Center of the Circumscribed Circle
      --- equidistant from each vertex.)
The circumcenter of an obtuse $\triangle$ is outside of the triangle.

e.g.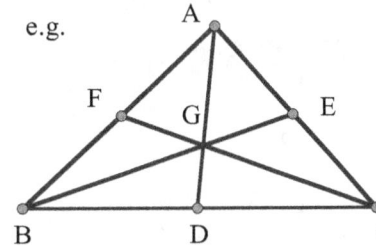

G is the centroid.
Then:
DG = x
AG = 2x
AD = 3x

4. Isosceles Triangle

Definition: A triangle that has two congruent sides.

(1) The base angles of an isosceles triangle are congruent.
(2) If two angles of a triangle are congruent, then their opposite sides are congruent.
(3) To the base of an isosceles triangle, the median, altitude, angle bisector, and perpendicular bisector coincide.

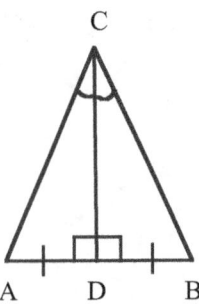

$\triangle ABC$ is isosceles with $\overline{AC} \cong \overline{BC}$,
$\overline{CD}$ is the median, altitude, angle bisector, and perpendicular bisector.

5. Right Triangle

Pythagorean Theorem

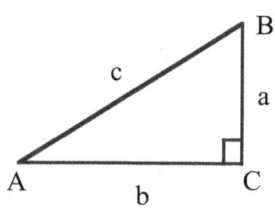

∠C is a right angle
a and b are legs
c is the hypotenuse

$$a^2 + b^2 = c^2$$

Pythagorean Triples:
    3, 4, 5;   6, 8, 10; 9, 12, 15 etc.
    5, 12, 13; 10, 24, 26 etc.

e.g. The ratio of two legs are 3:4 and the hypotenuse is 15.

    Find the lengths of the two legs:
        $(3n)^2 + (4n)^2 = 15^2$
          $9n^2 + 16n^2 = 15^2$
             $25n^2 = 225$
               $n^2 = 9$
                $n = 3$ (n = - 3 rejected)
   $3n = 3•3 = 9$ and $4n = 4•3 = 12$
   The lengths of the two legs are 9 and 12.

Special Right Triangles

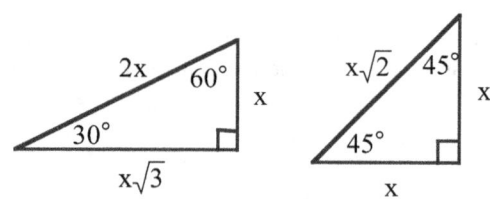

6. Congruent Triangles
SSS ≅
SAS ≅
ASA ≅
AAS ≅   (No SSA ≅)
HL ≅ (for right triangles only)
CPCTC: Corresponding Parts of Congruent Triangles
are Congruent.

---

e.g. $\overline{AB}$ and $\overline{CD}$ bisect each other at point E.
     Prove: △ACE ≅ △BDE

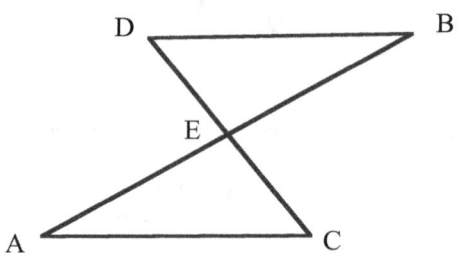

1. $\overline{AB}$ and $\overline{CD}$ bisect each other at point E	1. Given
2. $\overline{AE}$ ≅ $\overline{BE}$ and $\overline{CE}$ ≅ $\overline{DE}$	2. A bisector divides a segment into two congruent segments
3. ∠AEC ≅ ∠BED	3. Vertical angles are congruent
4. △ACE ≅ △BDE	4. SAS ≅

e.g. $\overline{CA} \perp \overline{AB}$, $\overline{DB} \perp \overline{AB}$, $\overline{AD}$ ≅ $\overline{BC}$
     Prove: △ABC ≅ △BAD

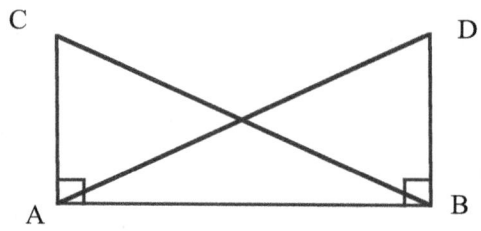

1. $\overline{CA} \perp \overline{AB}$, $\overline{DB} \perp \overline{AB}$	1. Given
2. ∠CAB and ∠DBA are right angles	2. ⊥ lines form right angles
3. △ABC and △BAD are right △s	3. Def. of the right △
4. $\overline{AD}$ ≅ $\overline{BC}$	4. Given
5. $\overline{AB}$ ≅ $\overline{AB}$	5. Reflexive property
6. △ABC ≅ △BAD	6. HL ≅

7. Ratios, Proportions, and Similar Triangles

(1) Ratios and Proportions

If two ratios are equal, they are in proportion.

$$\frac{a}{b} = \frac{c}{d} \quad \text{or} \quad a \cdot d = b \cdot c$$

In a proportion, the product of the means is equal to the product of the extremes.

(2) Similar Triangles

AA ~ (most often used for proof)
SAS ~
SSS ~

If two triangles are similar, then their corresponding angles are congruent and their corresponding sides are in proportion.

In two similar triangles, the ratio of the perimeters is equal to the ratio of the sides.

In two similar triangles, the ratio of the areas is equal to the square of the ratio of the sides.

e.g If $\triangle ABC \sim \triangle A'B'C'$ and $\dfrac{AB}{A'B'} = \dfrac{2}{1}$

then $\dfrac{P}{P'} = \dfrac{2}{1}$ and $\dfrac{\text{Area}}{\text{Area}'} = \dfrac{4}{1}$

Theorem

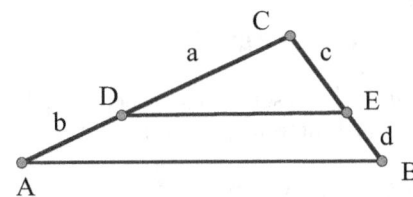

If $\overline{DE} \parallel \overline{AB}$, then $\dfrac{a}{b} = \dfrac{c}{d}$

Midsegment Theorem

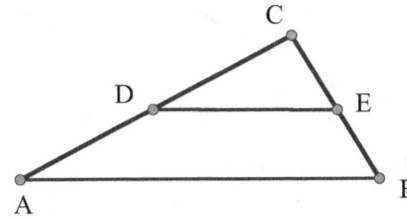

If D and E are midpoints of $\overline{AC}$ and $\overline{BC}$, then the midsegment $\overline{DE}$ is parallel to $\overline{AB}$ and is half of $\overline{AB}$.

e.g.

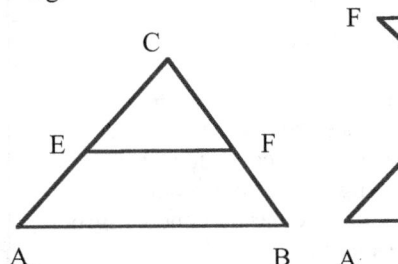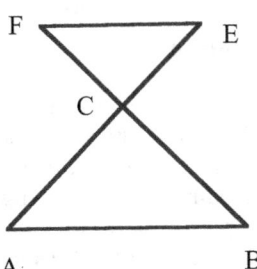

If $\overline{EF} \parallel \overline{AB}$,
then $\triangle EFC \sim \triangle ABC$ AA~

then $\dfrac{EC}{AC} = \dfrac{FC}{BC} = \dfrac{EF}{AB}$ Corresponding sides in proportion

e.g.

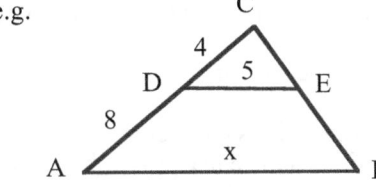

If $\overline{AB} \parallel \overline{DE}$

then $\dfrac{5}{x} = \dfrac{4}{4+8}$, $x = 15$

(Note: $\dfrac{5}{x} \neq \dfrac{4}{8}$)

(3) Proportions in the Right Triangle

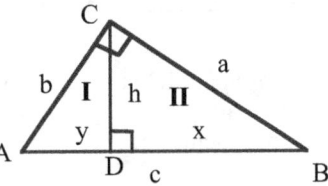

Right Triangle Altitude Theorem

(a). $\triangle I \sim \triangle II \sim \triangle ABC$

(b).   The altitude to the hypotenuse is the geometric mean of the two segments of the hypotenuse.
$$h^2 = x\,y \;;$$

(c).   Each leg is the geometric mean of its projection on the hypotenuse and the whole hypotenuse.
$$a^2 = x\,(x + y) = xc \;;$$
$$b^2 = y\,(x + y) = yc$$

VI. POLYGONS

1. Quadrilateral: a 4-sided polygon.

2. To Prove a Parallelogram:
  2 pairs of opposite sides are parallel;
  2 pairs of opposite sides are congruent;
  2 pairs of opposite angles are congruent;
  1 pair of opposite sides are parallel and congruent;
  Diagonals bisect each other.
Rhombus: All the properties of the parallelogram;
  4 sides are congruent;
  Diagonals are perpendicular;
  Diagonals bisect the interior angles.
Rectangle: All the properties of the parallelogram;
  4 right angles;
  Diagonals are congruent.
Square: All the properties of the rhombus and the rectangle.

3. Trapezoid: one and only one pair of opposite sides
  are parallel.
  The median of a trapezoid is parallel to the bases.
  The length of the median is equal to one-half the
  sum of the lengths of the bases.

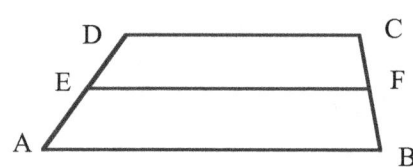

$\overline{EF}$ is the median: $\overline{EF} \parallel \overline{AB}$, $\overline{EF} \parallel \overline{CD}$

$$EF = \frac{AB + CD}{2}$$

Isosceles Trapezoid: the nonparallel sides are congruent.
  Base angles are congruent.
  Diagonals are congruent.

4. Polygons

Sum of the exterior angles $= 360°$

Exterior angle of a regular polygon $= \dfrac{360°}{n}$

Sum of the interior angles $= n \cdot 180° - 360°$
$$= (n - 2) \cdot 180°$$

Interior angle of a regular polygon $= 180° - \dfrac{360°}{n}$

Square (4 sides), Pentagon (5 sides), Hexagon (6 sides),
Octagon (8 sides).

VII. COORDINATE GEOMETRY

1. Slope, Midpoint, Distance, and Centroid

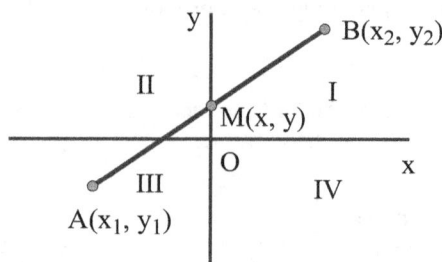

Coordinate Plane has four Quadrants I, II, III, and IV.

slope $m = \dfrac{y_2 - y_1}{x_2 - x_1}$

midpoint $M(\bar{x}, \bar{y}) = M(\dfrac{x_1 + x_2}{2}, \dfrac{y_1 + y_2}{2})$

distance $d = \sqrt{(x_2 - x_1)^2 + (y_2 - y_1)^2}$

e.g. $\overline{AB}$ has midpoint $M(1,4)$ and one end $B(3,5)$.
Find the coordinates of the other end A.

$$1 = \frac{x_1 + 3}{2} \quad , \quad 4 = \frac{y_1 + 5}{2}$$

  Solve for x_1 and y_1. $A(x_1, y_1) = A(-1, 3)$
e.g. The **Centroid** of a triangle is

$$(\frac{x_1 + x_2 + x_3}{3}, \frac{y_1 + y_2 + y_3}{3})$$

2. Linear Function (First Degree)

A straight line can be represented as a linear function;
The graph of a linear function is a straight line.

Slope-Intercept Form: $y = mx + b$
where m is the slope and b is the y - intercept.

Point-Slope Form: $y - y_1 = m(x - x_1)$

e.g.

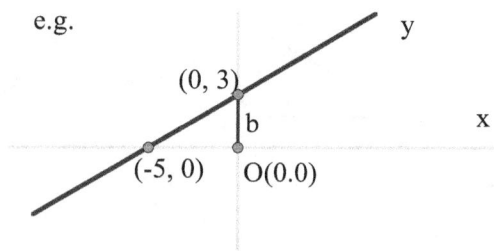

$b = 3, \quad m = \dfrac{3 - 0}{0 - (-5)} = \dfrac{3}{5}, \quad y = \dfrac{3}{5}x + 3$

Special cases:
Direct Variation: when b = 0,
the line passing through the Origin
$$y = mx$$
Vertical line: x = a
Horizontal line: y = b

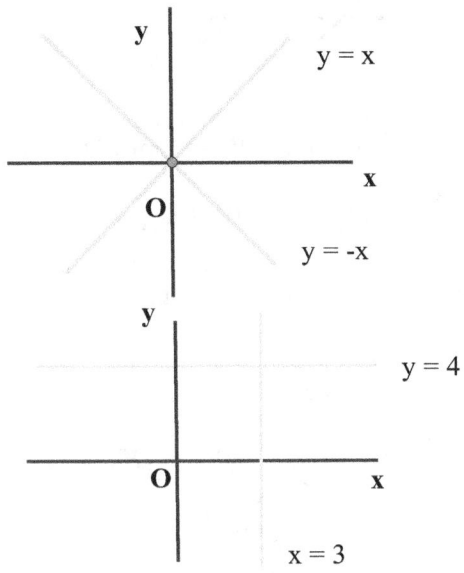

Two parallel lines have the same slope ($m_1 = m_2$);

Two perpendicular lines: $m_2 = -\dfrac{1}{m_1}$ or $m_1 \bullet m_2 = -1$;

The slope of a horizontal line is zero (m = 0);
The slope of a vertical line is undefined.

e.g. Find the slope and y-intercept of 3x - 2y = 12.
Write the equation in slope and y-intercept form:
$$y = \frac{3}{2}x - 6 \quad , \quad \text{slope } m = \frac{3}{2} \text{ and y-intercept } b = -6$$

e.g. Write the equation of a line through (3, -2) and (6, 4).
First find the slope m = $\dfrac{4 - (-2)}{6 - 3} = \dfrac{6}{3} = 2$
y = 2x + b ,   replace x by 6 and y by 4
4 = 2•6 + b   solve for b = -8
We have the equation of the line y = 2x - 8

e.g. Write the equation of a line passing through the origin
and perpendicular to the line y = 2x + 3.
Since the line passing through the origin: y = mx   (b = 0)
$$m_2 = -\frac{1}{m_1} = -\frac{1}{2} \quad , \qquad \text{Solution: } \quad y = -\frac{1}{2}x$$

3. Coordinate Geometric Proof

(1).To prove a parallelogram:
Method 1: (Slope formula)
Two pairs of opposite sides are parallel - the same slope;
Method 2: (Midpoint formula)
Diagonals having the same midpoint bisect each other.
(2). To prove a rhombus:
(Distance formula)
4 sides have the same length;
(3). To prove a rectangle:
(Slope formula)
Opposite sides are parallel, adjacent sides are
perpendicular;
(4). To prove a trapezoid:
(Slope formula)
One pair of the opposite sides are parallel - the same slope;
and the other pair of the opposite sides are not parallel -
different slopes.

e.g. The quadrilateral ABCD has vertices A(-5, -2),
B(-5, 3), C(4, 6), and D(7, 2). Prove by coordinate
geometry that quadrilateral ABCD is an isosceles
trapezoid

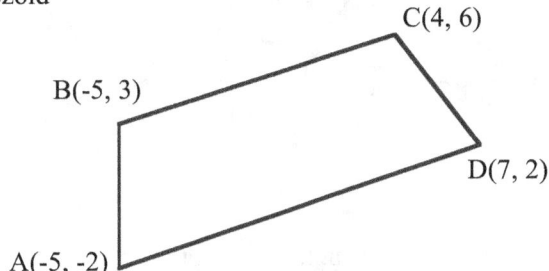

Prove:

Slope of $\overline{AD}$ = $\dfrac{2 - (-2)}{7 - (-5)} = \dfrac{4}{12} = \dfrac{1}{3}$

Slope of $\overline{BC}$ = $\dfrac{6 - 3}{4 - (-5)} = \dfrac{3}{9} = \dfrac{1}{3}$

Slope of $\overline{AD}$ = Slope of $\overline{BC}$ $\overline{AD} \parallel \overline{BC}$

Slope of $\overline{AB}$ = $\dfrac{3 - (-2)}{-5 - (-5)} = \dfrac{5}{0}$ (vertical line)

Slope of $\overline{CD}$ = $\dfrac{2 - 6}{7 - 4} = \dfrac{-4}{3}$

Slope of $\overline{AB}$ ≠ Slope of $\overline{CD}$ $\overline{AB}$ is not $\parallel \overline{CD}$
Therefore ABCD is a trapezoid.

$AB = \sqrt{[-5 - (-5)]^2 + [3 - (-2)]^2} = 5$
$CD = \sqrt{(7 - 4)^2 + (2 - 6)^2} = 5$
AB = CD
Therefore ABCD is an isosceles trapezoid.

VIII. CIRCLE

1. Angles of a Circle
The degree measure of a circle is 360°.
The degree measure of a semicircle is 180°.
The degree measure of an arc is equal to the measure of the central angle that intercepts the arc.

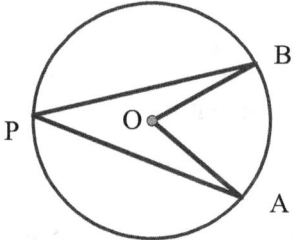

Central angle $m\angle AOB = m\overarc{AB}$

Inscribed angle $m\angle APB = \dfrac{1}{2}m\overarc{AB}$

Inscribed angle of a semicircle is a right angle, vice versa.

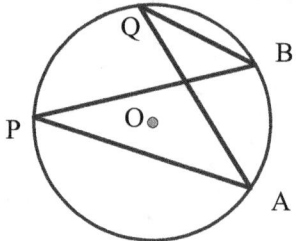

If two inscribed angles intercept the same arc, they are congruent.
$$\angle P \cong \angle Q$$

Congruent arcs have congruent chords, vice versa.

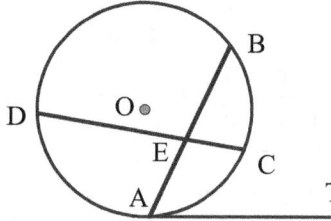

Chord - tangent angle: $m\angle BAT = \dfrac{1}{2}m\overarc{BCA}$

Chord - chord angle: $m\angle BEC = \dfrac{1}{2}(m\overarc{BC} + m\overarc{AD})$;

$$m\angle AEC = \dfrac{1}{2}(m\overarc{AC} + m\overarc{BD})$$

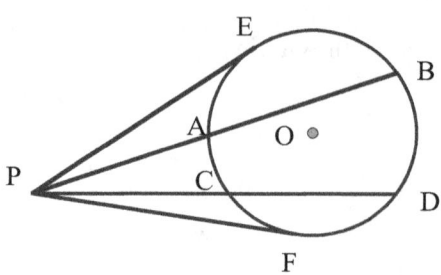

Tangent - tangent angle $m\angle EPF = \dfrac{1}{2}(m\overarc{EBDF} - m\overarc{EF})$

Tangent - secant angle $m\angle EPB = \dfrac{1}{2}(m\overarc{EB} - m\overarc{EA})$

Secant - secant angle $m\angle BPD = \dfrac{1}{2}(m\overarc{BD} - m\overarc{AC})$

e.g.

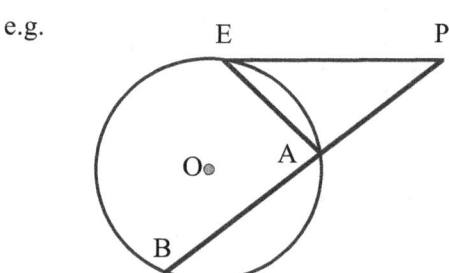

Given: $\overline{PE}$ is tangent.
$m\overarc{EA} : m\overarc{AB} : m\overarc{BE} = 2 : 3 : 4$

Find: $m\angle P$, $m\angle PEA$, $m\angle PAE$

$m\overarc{EA} : m\overarc{AB} : m\overarc{BE} = 2x : 3x : 4x$
$2x + 3x + 4x = 360 \qquad x = 40$

$m\overarc{EA} = 80$, $m\overarc{AB} = 120$, $m\overarc{BE} = 160$

$m\angle P = \dfrac{1}{2}(m\overarc{BE} - m\overarc{EA}) = 40$

$m\angle PEA = \dfrac{1}{2}m\overarc{EA} = 40$

$m\angle PAE = 180 - 40 - 40 = 100$

2. Segments of a Circle

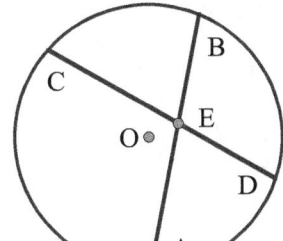

$$AE \cdot EB = CE \cdot ED$$

$$\frac{AE}{CE} = \frac{ED}{EB}$$

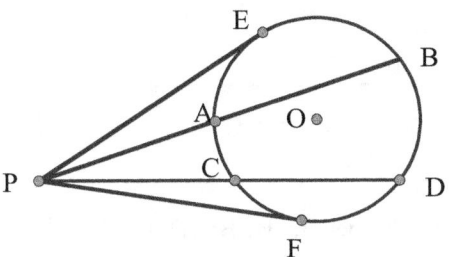

$\overline{PE}$, $\overline{PF}$ are tangent segments. $\overline{PE} \cong \overline{PF}$

$PE^2 = PA \cdot (PA + AB) = PC \cdot (PC + CD)$

$$\frac{PA}{PE} = \frac{PE}{PB} \qquad \frac{PC}{PE} = \frac{PE}{PD} \qquad \frac{PA}{PC} = \frac{PD}{PB}$$

3. Theorems

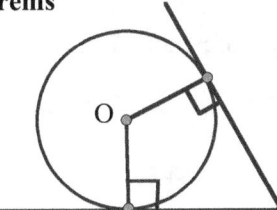

A tangent to a circle is perpendicular to the radius at its point of intersection.

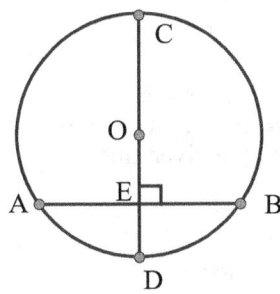

A diameter perpendicular to a chord bisects the chord and its arcs.

If $\overline{AB} \perp \overline{CD}$, then $\overline{AE} \cong \overline{BE}$ and $\overparen{AD} \cong \overparen{BD}$, $\overparen{AC} \cong \overparen{BC}$.

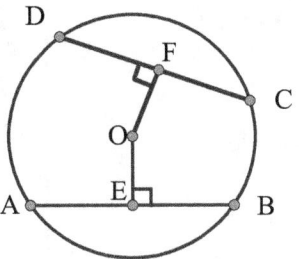

If two chords of a circle are congruent, then they are equidistant from the center of the circle, vice versa.

If $\overline{AB} \cong \overline{CD}$, then $\overline{OE} = \overline{OF}$;   or

If $OE = OF$, then $\overline{AB} \cong \overline{CD}$

If $OE < OF$, then $AB > CD$

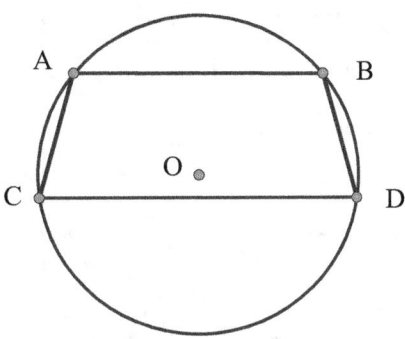

In a circle, parallel chords intercept congruent arcs between them.

If $\overline{AB} \parallel \overline{CD}$, then $\overparen{AC} \cong \overparen{BD}$

In a circle, congruent arcs have congruent chords, vice versa.

If $\overparen{AC} \cong \overparen{BD}$, then $\overline{AC} \cong \overline{BD}$

4. Common Tangents of Two Circles:

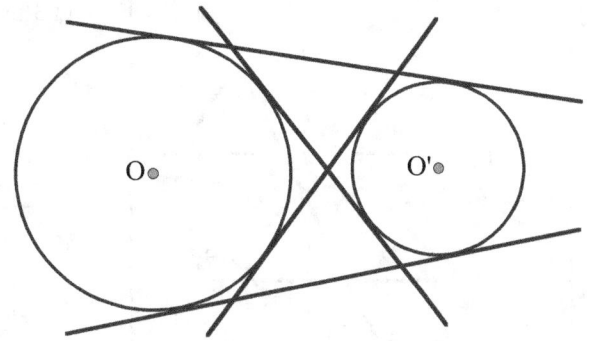

IX. CONSTRUCTIONS AND LOCI

1. Three Types of Constructions

(1) Copy a line segment or an angle.

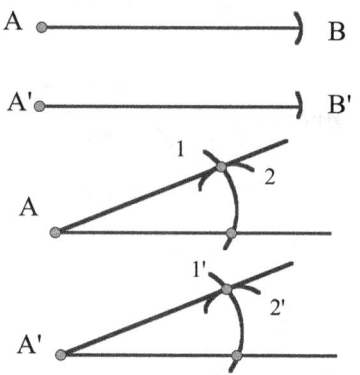

(2) Bisect a line segment or an angle.

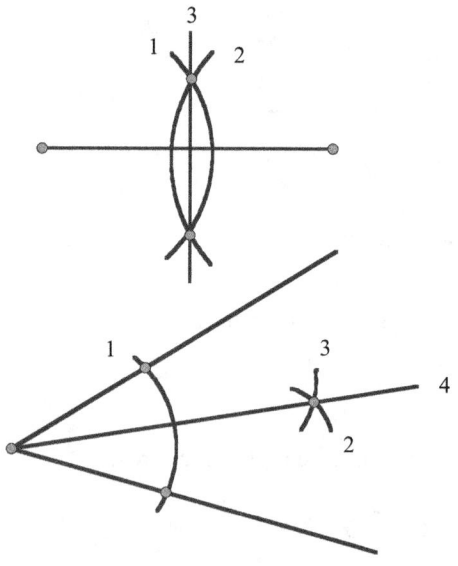

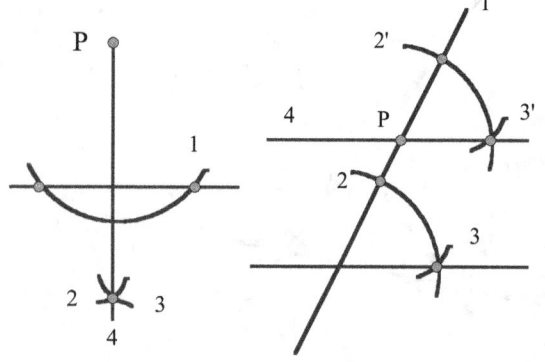

(3) Through a point draw a line ⊥ or ∥ to a given line.

2. Five Fundamental Loci

(1) The locus of points equidistant from a given point.
(2) The locus of points equidistant from two given points.
(3) The locus of points equidistant from two sides of a given angle.
(4) The locus of points equidistant from a given line.
(5) The locus of points equidistant from two given parallel lines.

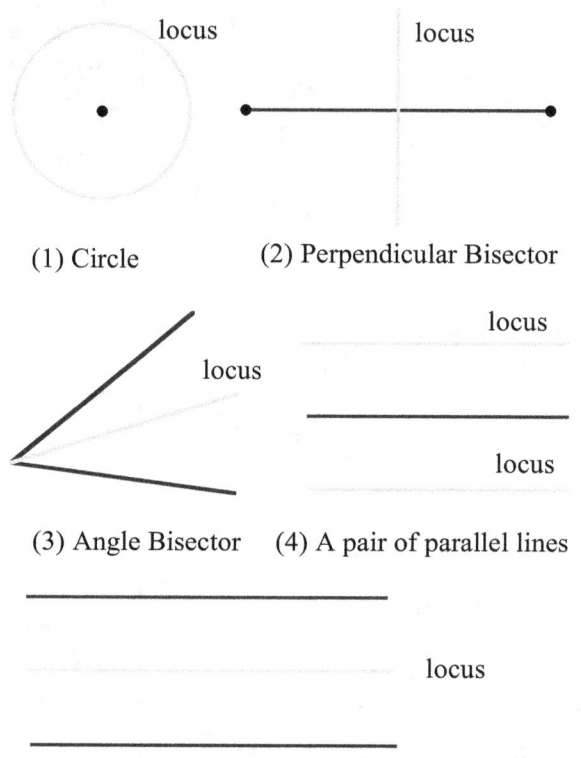

(1) Circle   (2) Perpendicular Bisector

(3) Angle Bisector   (4) A pair of parallel lines

(5) A parallel line midway between the given lines

3. Compound Loci

Find the points of intersection of different loci.
e.g. How many points are 2 units from a given line and 3 units from a given point on the given line ?

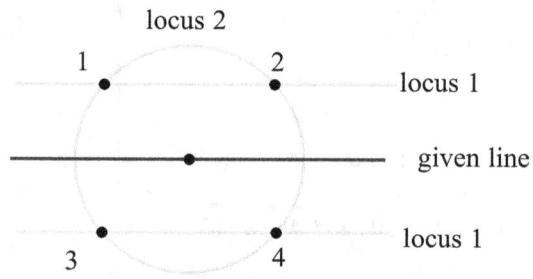

There are four points of intersection.

4. Equations of Loci

(1) The center-radius equation of a circle with radius r and center (h, k)

$$(x - h)^2 + (y - k)^2 = r^2$$

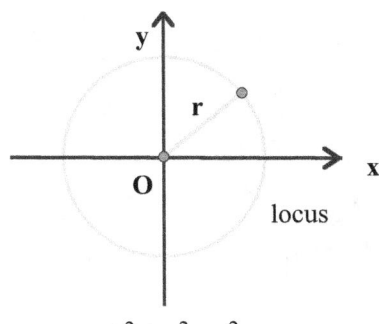

$$x^2 + y^2 = r^2$$

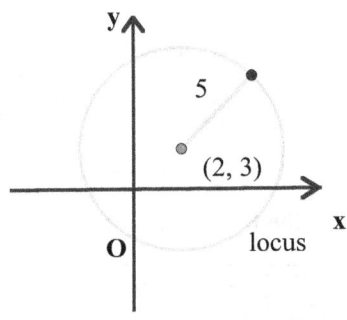

$$(x - 2)^2 + (y - 3)^2 = 5^2$$

(2) The equation of the perpendicular bisector

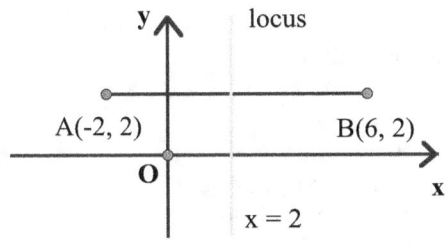

x = 2 is the equation of the perpendicular bisector of $\overline{AB}$.

Find the equation of the locus of points equidistant from points A(-2, 2) and B(4, -2).

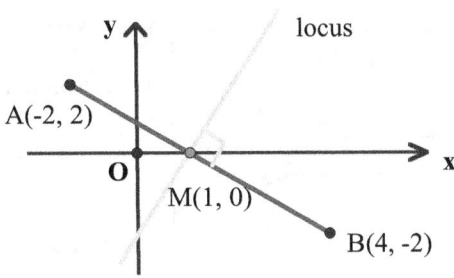

Find the midpoint of $\overline{AB}$:

$$M(\bar{x}, \bar{y}) = M(\frac{-2 + 4}{2}, \frac{2 + (-2)}{2})$$
$$= M(1, 0)$$

Find the slope of $\overline{AB}$:

$$m_1 = \frac{-2 - 2}{4 - (-2)} = \frac{-4}{6} = -\frac{2}{3}$$

the slope of the perpendicular line:

$$m_2 = -\frac{1}{m_1} = \frac{3}{2}$$

the equation of the perpendicular bisector:

the slope is $\frac{3}{2}$ and passing through midpoint (1, 0)

the point-slope form: $y - y_1 = m(x - x_1)$

$$y - 0 = \frac{3}{2}(x - 1), \quad \text{or}$$

slope-intercept form: $y = \frac{3}{2}x - \frac{3}{2}$

(3) The equation of the angle bisector.

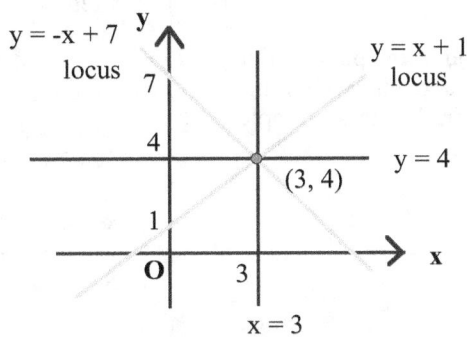

y = x + 1 and y = -x + 7 are the loci of the points equidistant from lines x = 3 and y = 4.
Hint: m = ± 1 and passing through point (3, 4)

(4) The equations of the locus of points equidistant from a given line.

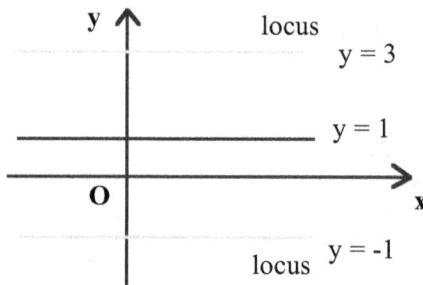

$y = -1$ and $y = 3$ are the equations of the parallel lines 2 units from the given line $y = 1$

(5) The equation of the locus of points equidistant from two given parallel lines.

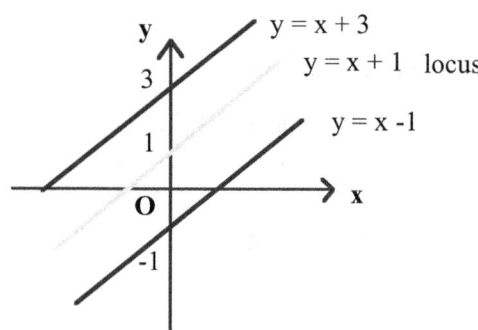

The y-intercept of the locus is equal to

$$\frac{-1 + 3}{2} = 1$$

$y = x + 1$ is the equation of the parallel line midway between the two given parallel lines. They have the same slope 1 in this example.

(6) The points of intersection of compound loci:

Find the solution to the quadratic-linear system of equations.

e.g. Solve the quadratic-linear system of equations
$$y = x^2 + 1$$
$$y = x + 3$$

Method 1: Solve graphically

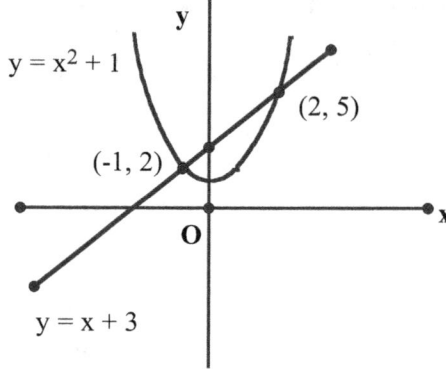

To graph the parabola $y = x^2 + 1$, properly choose 7 points around the turning point:

x	-3	-2	-1	0	1	2	3
y	10	5	2	1	2	5	10

The points of intersection (-1, 2) and (2, 5) are the solution to the quadratic-linear system of equations.

Method 2: Solve algebraically
$$y = x^2 + 1$$
$$y = x + 3$$
$x^2 + 1 = x + 3$ substitution
$x^2 - x - 2 = 0$ make the right side 0
$(x - 2)(x + 1) = 0$ factor the left side
$x - 2 = 0$ or $x + 1 = 0$
$x = 2$ or $x = -1$
$y = x + 3 = 5$ $y = x + 3 = 2$
Solution: $\{(2, 5), (-1, 2)\}$

X. TRANSFORMATION

1. Transformation Rules

(1). Line Reflection:
$$P(x, y) \underline{\quad rx\text{-axis} \quad} P'(x, -y)$$
$$P(x, y) \underline{\quad ry\text{-axis} \quad} P'(-x, y)$$
$$P(x, y) \underline{\quad ry = x \quad} P'(y, x)$$
$$P(x, y) \underline{\quad ry = -x \quad} P'(-y, -x)$$

(2). Point Reflection:
$$P(x, y) \underline{\quad ro \quad} P'(-x, -y)$$

(3). Translation:
$$P(x, y) \underline{\quad T_{a, b} \quad} P'(x + a, y + b)$$

(4). Rotation:
$$P(x, y) \underline{\quad R\ 90° \quad} P'(-y, x)$$
$$P(x, y) \underline{\quad R\ 180° \quad} P'(-x, -y)$$
$$P(x, y) \underline{\quad R\ -90° \quad} P'(y, -x)$$

(5). Dilation:
$$P(x, y) \underline{\quad D\ k \quad} P'(kx, ky)$$

Only dilation enlarges or reduces the size of the image, which is similar to the original.
The image of other transformations is congruent to the original.

It is not practical to memorize these rules. One should be able to derive these rules through the drawings.

e.g. Find the rules for ry-axis and R -90°.

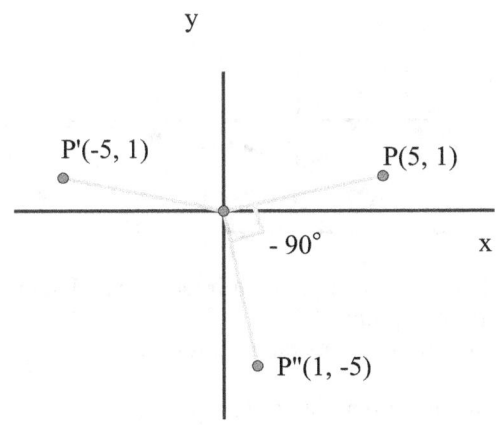

$$P(5, 1) \underline{\quad ry\text{-axis} \quad} P'(-5, 1)$$
$$P(x, y) \underline{\quad ry\text{-axis} \quad} P'(-x, y)$$
$$P(5, 1) \underline{\quad R\ -90° \quad} P''(1, -5)$$
$$P(x, y) \underline{\quad R\ -90° \quad} P''(y, -x)$$

e.g

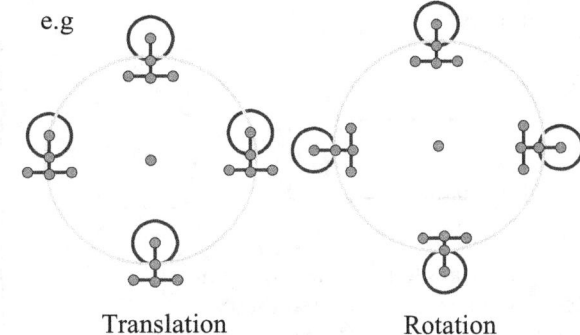

Translation   Rotation

2. Composition of Transformations

e.g. rx-axis ∘ ry-axis (x, y) = rx-axis (-x, y) = (-x, y)
    We can see   rx-axis ∘ ry-axis = ro

e.g. $D_4 \circ T_{3, 0}(x, y) = D_4(x + 3, y) = (4x + 12, 4y)$

or $(x, y) \xrightarrow{\quad T_{3, 0} \quad} (x + 3, y) \xrightarrow{\quad D_4 \quad} (4x + 12, 4y)$

Glide Reflection:

Glide Reflection is a special composition of reflections and translations: $T_{a, 0} \circ r_{x\text{-axis}}$

e.g. Find the image of P(1, 2) under the glide reflection of $T_{2, 0} \circ r_{x\text{-axis}}$

$$P(1, 2) \underline{\quad rx\text{-axis} \quad} P'(1, -2) \underline{\quad T_{2, 0} \quad} P''(3, -2)$$

3. Functions under a Transformation

$$y = f(x) \underline{\quad T_{a, b} \quad} y = f(x - a) + b$$

$$y = f(x) \underline{\quad rx\text{-axis} \quad} y = -f(x)$$

The transformation rules for functions are different from the transformation rules for images

e.g. $y = x^2 \underline{\quad T_{5, 2} \quad} y = (x - 5)^2 + 2$
     $y = x^2 \underline{\quad rx\text{-axis} \quad} y = -x^2$

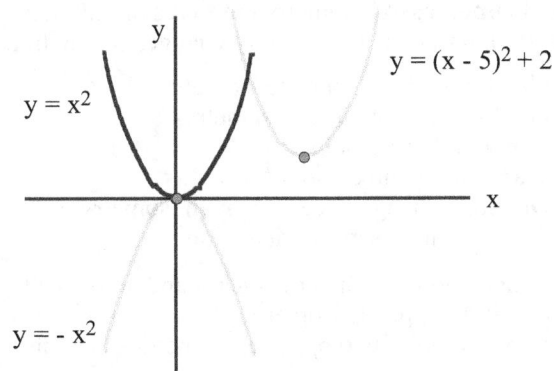

4. Symmetry

(1). Line Symmetry

e.g.

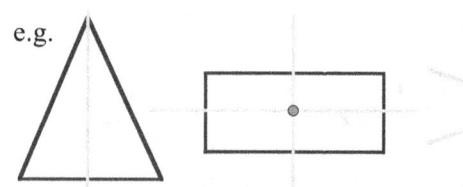

Isosceles triangle has 1 axis of symmetry;
Rectangle has 2 axes of symmetry;
Regular pentagon has 5 axes of symmetry.

(2). Point Symmetry

e.g.

(3). Rotational Symmetry

e.g.

 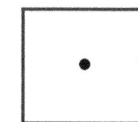 

Equilateral triangle has 120° rotational symmetry;
Square has 90° rotational symmetry;
Rectangle has 180° rotational symmetry;
Regular pentagon has 72° rotational symmetry;
Regular hexagon has 60° rotational symmetry.

5. Isometry and Orientation

Isometry: A transformation that preserves the distance.
Only dilation is not isometry.

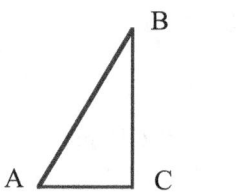 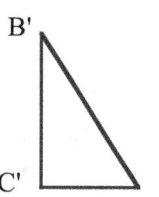

From A ---> B ---> C is **clockwise orientation**.
From A' ---> B' ---> C' is **counterclockwise orientation**.

Direct Isometry: An isometry preserves orientation;
Opposite Isometry: An isometry changes orientation.

e.g.   Line Reflection: Opposite Isometry
     Point Reflection: Direct Isometry
     Rotation: Direct Isometry
     Translation: Direct Isometry
     Dilation: Changes size --- Not an isometry,
           but preserves orientation

The composition of a direct isometry and an opposite
isometry is an opposite isometry.
The composition of two opposite isometries is a direct
isometry.

XI. SOLID GEOMETRY

1. To determine a plane

(1). Three noncollinear points determine a plane.

(2). Two intersecting lines determine a plane.

(3). Two parallel lines determine a plane.

(4). Skew lines are neither parallel nor intersecting,
they are not in a same plane --- not coplanar.

e.g. In a triangular right prism

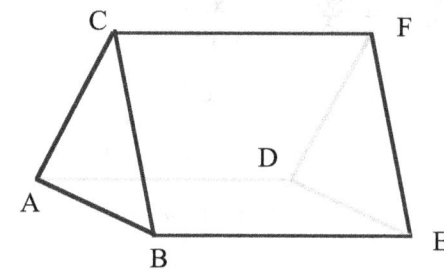

$\overline{AB} \parallel \overline{DE}$, $\overline{BC} \parallel \overline{EF}$, $\overline{CA} \parallel \overline{FD}$;
$\overline{AD} \parallel \overline{BE} \parallel \overline{CF}$;
$\overline{AB}$ and $\overline{CF}$, $\overline{BC}$ and $\overline{AD}$, $\overline{CA}$ and $\overline{BE}$,
$\overline{AB}$ and $\overline{FD}$, $\overline{AB}$ and $\overline{FE}$ etc. are pairs of skew lines.

2. A line perpendicular to a plane

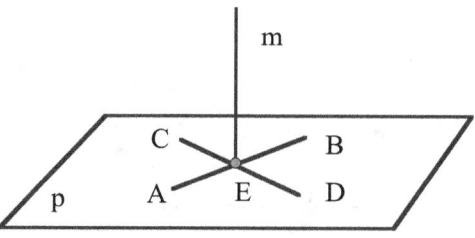

(1). If a line is not in a plane, it intersects a plane in
exactly one point.
e.g. line m intersects plane p at point E

(2). If a line is perpendicular to a plane, it is
perpendicular to each line in the plane through
the point of intersection.
e.g. $m \perp \overleftrightarrow{AB}$, $m \perp \overleftrightarrow{CD}$

(3). Through a given point (on the plane or not on
the plane), there is one and only one line
perpendicular to the given plane.

(4). Two lines perpendicular to a same plane are parallel and coplanar.

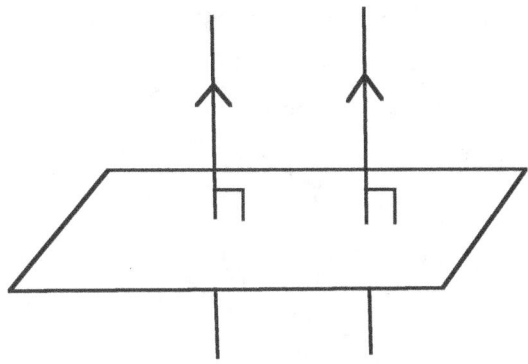

3. Perpendicular Planes

(1). Two perpendicular planes intersect to form a right dihedral angle.

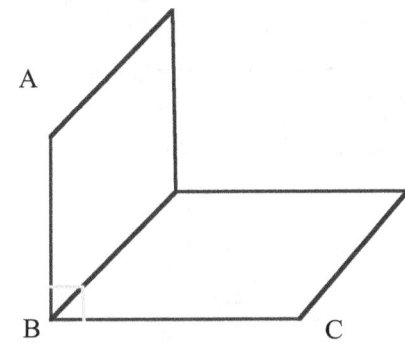

(2). If plane a and plane b are both perpendicular to plane c , then their line of intersection m is perpendicular to plane c.

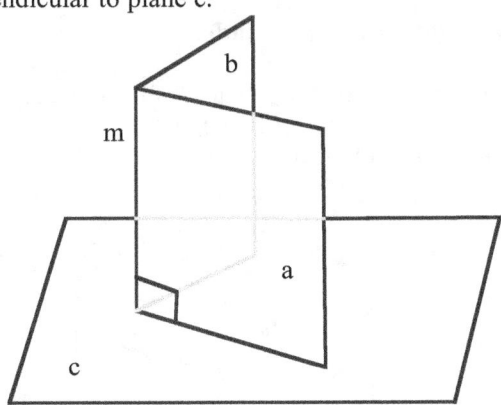

(3). If a plane contains a line perpendicular to another plane, then these two planes are perpendicular.

4. Parallel Planes

(1). Parallel planes are everywhere equidistant.

(2). If two planes are perpendicular to a same line, then they are parallel.

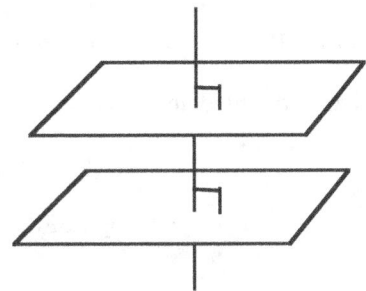

(3). If a plane intersects two parallel planes, then the intersection is two parallel lines.

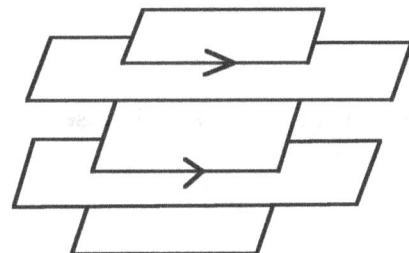

XII. INDIRECT PROOF

(1) Assume that the opposite of the conclusion is true.
(2) Show that the assumption contradicts a known fact.
(3) Since the assumption is false, the conclusion is true.

e.g.

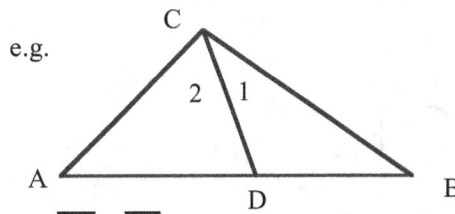

Given: $\overline{AC} \cong \overline{BC}$ and $\angle 1 \neq \angle 2$
Prove: $\overline{CD}$ is not a median

(1) $\overline{AC} \cong \overline{BC}$, $\angle 1 \neq \angle 2$ Given
(2) $\overline{CD}$ is a median   Assumed
(3) D is the midpoint   Def. of median
(4) $\overline{AD} \cong \overline{BD}$ Def. of midpoint
(5) $\overline{CD} \cong \overline{CD}$ Reflexive Property
(6) $\triangle ACD \cong \triangle BCD$ S.S.S $\cong$
(7) $\angle 1 \cong \angle 2$ CPCTC
(8) $\overline{CD}$ is not a median   Contradiction in (7) and (1)

XIII. GEOMETRIC MEASUREMENTS

1 yd = 3 ft ,   1 ft = 12 in ,   1 mile = 5280 ft
1 m = 100 cm , 1 m = 1000 mm

1. Circle

   Circumference $C = 2\pi r = \pi d$ r: radius   d: diameter
      Area $A = \pi r^2$
e.g. When r is doubled, C is doubled and A increases 4 times.

2. Square

   Perimeter $P = 4s$ s: length of the side
      Area $A = s^2$

3. Rectangle

   Perimeter $P = 2l + 2w$ l: length   w: width
   Area $A = l \cdot w$

4. Parallelogram

   Perimeter P = sum of the 4 sides   b: base   h: height
   Area $A = b \cdot h$

5. Trapezoid

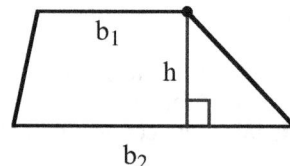

P = sum of 4 sides

$$A = \frac{b_1 + b_2}{2} \cdot h$$

6. Rhombus

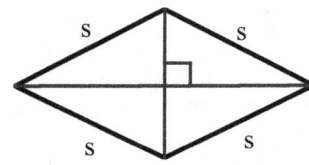

$P = 4s$

$$A = \frac{1}{2} \cdot d_1 \cdot d_2$$

d_1 and d_2 are diagonals

7. Triangle

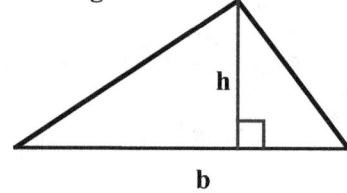

P = sum of 3 sides

$$A = \frac{1}{2} \cdot b \cdot h$$

8. Right Prism

Volume $V = Bh$ B: Base Area, h: height
Surface Area = 2 Base Areas + Lateral Areas

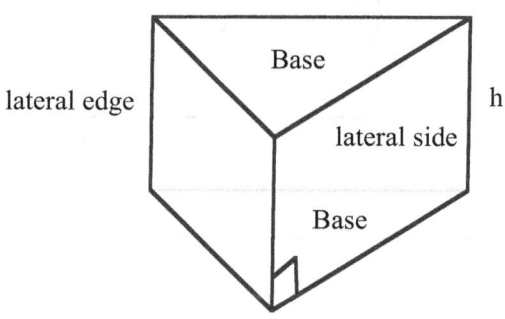

 All of the lateral edges are congruent and parallel.
Special cases:

Rectangular Prism
Volume $V = lwh$ l: length, w: width, h: height
Surface Area $SA = 2wl + 2wh + 2hl$

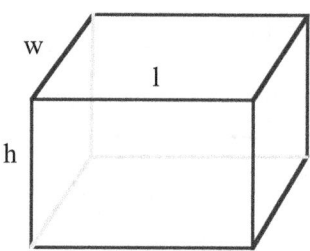

Cube
Volume $V = s^3$ s : length of the side
Surface Area $SA = 6s^2$

9. Right Circular Cylinder

Volume $V = Bh$ B: area of the circular base πr^2
                         h: height
Lateral Area $L = 2\pi rh$

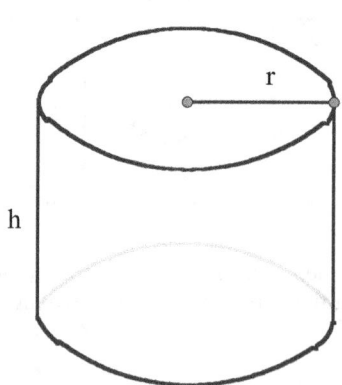

10. Pyramid

Volume $V = \dfrac{1}{3}Bh$ B: Base Area, h: height

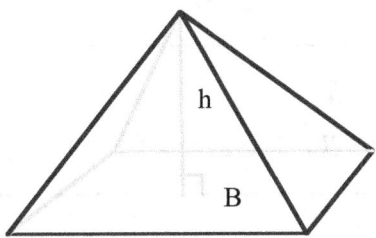

The base of a regular pyramid is a regular polygon. The lateral sides of a regular pyramid are congruent isosceles triangles.

11. Right Circular Cone

Volume $V = \dfrac{1}{3}Bh$ B: Base Area = πr^2, h: height

Lateral Area $L = \pi rl$ l: slant height

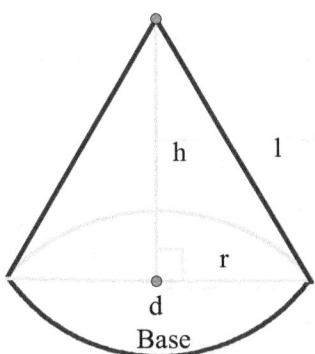

e.g.
A right circular cone has a diameter of 16 and a height of 18. Find the volume.
(a). express the answer in terms of π .
(b). express the answer to the nearest tenth

$r = \dfrac{d}{2} = \dfrac{16}{2} = 8$

$B = \pi r^2 = \pi \cdot 8^2 = 64\pi$

$V = \dfrac{1}{3}Bh = \dfrac{1}{3} \cdot 64\pi \cdot 18$

(a). $V = 384\pi$

(b). $V = 1206.4$

12. Sphere

Volume $V = \dfrac{4}{3}\pi r^3$

Surface Area $SA = 4\pi r^2$

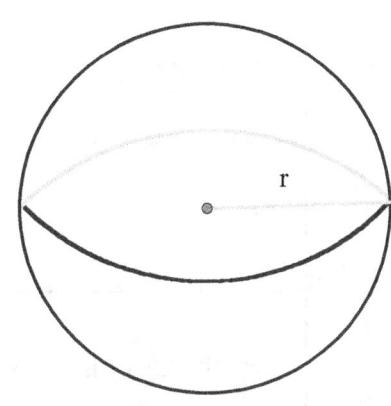

e.g.
A sphere has a diameter of 12 inches. Find the surface area of the sphere, to the nearest square inch.

$r = \dfrac{d}{2} = \dfrac{12}{2} = 6$

$SA = 4\pi r^2 = 4\pi \cdot 6^2 \approx 452$

Geometry Reference Sheet

Volume	Cylinder	$V = Bh$ where B is the area of the base
	Pyramid	$V = \frac{1}{3}Bh$ where B is the area of the base
	Right Circular Cone	$V = \frac{1}{3}Bh$ where B is the area of the base
	Sphere	$V = \frac{4}{3}\pi r^3$

Lateral Area (L)	Right Circular Cylinder	$L = 2\pi rh$
	Right Circular Cone	$L = \pi rl$ where l is the slant height

Surface Area	Sphere	$SA = 4\pi r^2$

I. LOGIC

1. What is the negation of the statement "The Sun is shining"?
(1) It is cloudy.   (3) It is not raining.
(2) It is daytime.   (4) The Sun is not shining.

2. What is the negation of the statement "Squares are parallelograms"?
(1) Parallelograms are squares.   (3) It is not the case that squares are parallelograms.
(2) Parallelograms are not squares. (4) It is not the case that parallelograms are squares.

3. What is the contrapositive of the statement, "If I am tall, then I will bump my head"?
(1) If I bump my head, then I am tall.   (3) If I am tall, then I will not bump my head.
(2) If I do not bump my head, then I am tall. (4) If I do not bump my head, then I am not tall.

4. What is the inverse of the statement "If two triangles are not similar, their corresponding angles are not congruent"?
(1) If two triangles are similar, their   (3) If two triangles are similar, their
     corresponding angles are not congruent.   corresponding angles are congruent.
(2) If corresponding angles of two triangles   (4) If corresponding angles of two triangles
     are not congruent, the triangles are not   are congruent, the triangles are similar.
     similar.

Show Work:

1. Given: Two is an even integer or three is an even integer.
Determine the truth value of this disjunction. Justify your answer.

2. Write a statement that is logically equivalent to the statement "If two sides of a triangle are congruent, the angles opposite those sides are congruent."
Identify the new statement as the converse, inverse, or contrapositive of the original statement.

IV. THEOREMS

1. The diagram below illustrates the construction of $\overleftrightarrow{PS}$ parallel to $\overleftrightarrow{RQ}$ through point P.

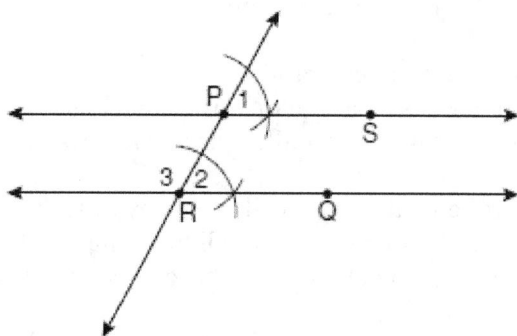

Which statement justifies this construction?

   (1) $m\angle 1 = m\angle 2$ (3) $\overline{PR} \cong \overline{RQ}$

   (2) $m\angle 1 = m\angle 3$ (4) $\overline{PS} \cong \overline{RQ}$

2. Based on the diagram below, which statement is true?

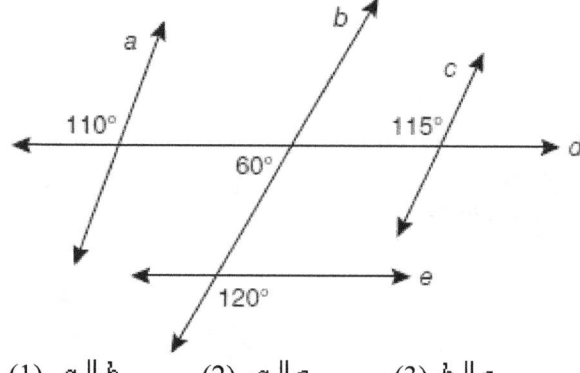

  (1) $a \parallel b$ (2) $a \parallel c$ (3) $b \parallel c$ (4) $d \parallel e$

V. TRIANGLES

1. Juliann plans on drawing △ABC, where the measure of ∠A can range from 50° to 60° and the measure of ∠B can range from 90° to 100°. Given these conditions, what is the correct range of measures possible for ∠C?
(1) 20° to 40°   (2) 30° to 50°   (3) 80° to 90°   (4) 120° to 130°

2. In an equilateral triangle, what is the difference between the sum of the exterior angles and the sum of the interior angles?
(1) 180°   (2) 120°   (3) 90°   (4) 60°

3. In △ABC, m∠$A = x$, m∠$B = 2x + 2$, and m∠$C = 3x + 4$. What is the value of x?
(1) 29   (2) 31   (3) 59   (4)61

4. In the diagram below, △ABC is shown with $\overline{AC}$ extended through point D.

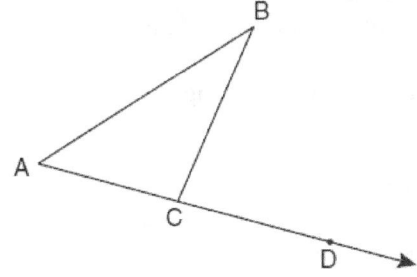

If m∠$BCD = 6x + 2$, m∠$BAC = 3x + 15$, and m∠$ABC = 2x - 1$, what is the value of x?

(1)12   (2) $14\dfrac{10}{11}$ (3) 16   (4) $18\dfrac{1}{9}$

5. In the diagram below of △ABC, D is a point on $\overline{AB}$, $AC = 7$, $AD = 6$, and $BC = 18$.

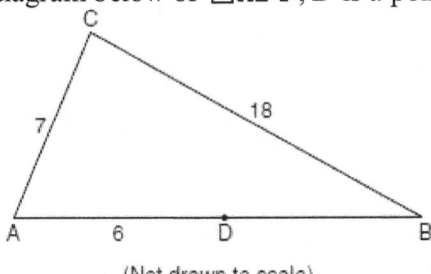

(Not drawn to scale)

The length of $\overline{DB}$ could be
(1) 5   (2) 12   (3) 19   (4) 25

6. In △ABC, m∠$A = 95$, m∠$B = 50$, and m∠$C = 35$. Which expression correctly relates the lengths of the sides of this triangle?
(1) $AB < BC < CA$ (2) $AB < AC < BC$ (3) $AC < BC < AB$ (4) $BC < AC < AB$

7. Side $\overline{PQ}$ of △PQR is extended through Q to point T. Which statement is *not* always true?
(1) m∠$RQT > $m∠$R$ (2) m∠$RQT > $m∠$P$ (3) m∠$RQT = $m∠$P + $m∠$R$ (4) m∠$RQT > $m∠$PQR$

8. Which set of numbers represents the lengths of the sides of a triangle?
(1) $\{5, 18, 13\}$ (2) $\{6, 17, 22\}$ (3) $\{16, 24, 7\}$ (4) $\{26, 8, 15\}$

9. Given $\triangle ABC$ with base $\overline{AFEDC}$, median $\overline{BF}$, altitude $\overline{BD}$, and $\overline{BE}$ bisects $\angle ABC$, which conclusion is valid?

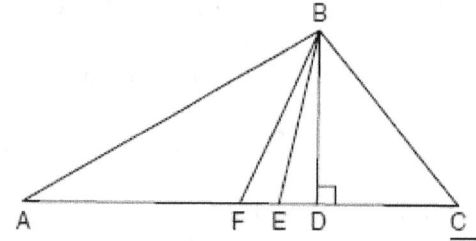

(1) $\angle FAB \cong \angle ABF$ (2) $\angle ABF \cong \angle CBD$ (3) $\overline{CE} \cong \overline{EA}$ (4) $\overline{CF} \cong \overline{FA}$

10. In which triangle do the three altitudes intersect outside the triangle?
(1) a right triangle   (2) an acute triangle   (3) an obtuse triangle   (4) an equilateral triangle

11. In the diagram of $\triangle ABC$ below, Jose found centroid P by constructing the three medians. He measured $\overline{CF}$ and found it to be 6 inches.

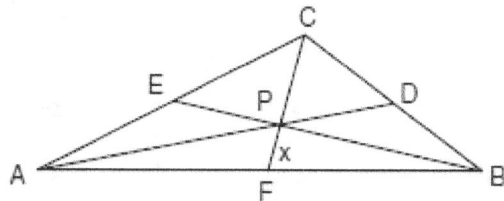

If $PF = x$, which equation can be used to find x?

(1) $x + x = 6$ (2) $2x + x = 6$ (3) $3x + 2x = 6$ (4) $x + \frac{2}{3}x = 6$

12. In the diagram of $\triangle ABC$ below, $\overline{AB} \cong \overline{AC}$. The measure of $\angle B$ is 40°.

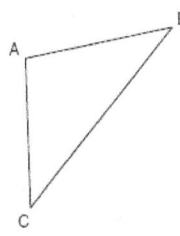

What is the measure of $\angle A$?
(1) 40°   (2) 50°   (3) 70°   (4) 100°

13. In $\triangle ABC$, $\overline{AB} \cong \overline{BC}$. An altitude is drawn from B to $\overline{AC}$ and intersects $\overline{AC}$ at D. Which conclusion is *not* always true?
(1) $\angle ABD \cong \angle CBD$ (2) $\angle BDA \cong \angle BDC$ (3) $\overline{AD} \cong \overline{BD}$ (4) $\overline{AD} \cong \overline{DC}$

14. In the diagram below of $\triangle ADB$, m$\angle BDA = 90$, $AD = 5\sqrt{2}$, and $AB = 2\sqrt{15}$.

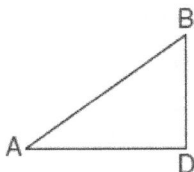

What is the length of $\overline{BD}$?
(1) $\sqrt{10}$ (2) $\sqrt{20}$ (3) $\sqrt{50}$ (4) $\sqrt{110}$

15. In the diagram of $\triangle ABC$ and $\triangle DEF$ below, $\overline{AB} \cong \overline{DE}$, $\angle A \cong \angle D$, and $\angle B \cong \angle E$.

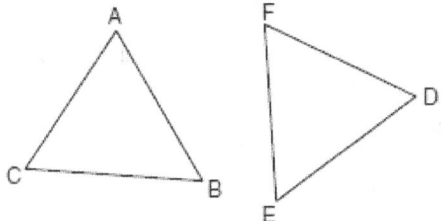

Which method can be used to prove $\triangle ABC \cong \triangle DEF$?
(1)   SSS   (2)   SAS   (3)   ASA   (4)   HL

16. In the diagram of trapezoid $ABCD$ below, diagonals $\overline{AC}$ and $\overline{BD}$ intersect at E and $\triangle ABC \cong \triangle DCB$.

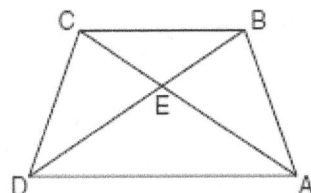

Which statement is true based on the given information?
(1) $\overline{AC} \cong \overline{BC}$ (2) $\overline{CD} \cong \overline{AD}$ (3) $\angle CDE \cong \angle BAD$ (4) $\angle CDB \cong \angle BAC$

17. The diagonal $\overline{AC}$ is drawn in parallelogram $ABCD$. Which method can *not* be used to prove that $\triangle ABC \cong \triangle CDA$?
(1)   SSS   (2)   SAS   (3)   SSA   (4)   ASA

18. Two triangles are similar, and the ratio of each pair of corresponding sides is 2 : 1. Which statement regarding the two triangles is *not* true?
(1) Their areas have a ratio of 4 : 1.   (2) Their altitudes have a ratio of 2 : 1.
(3) Their perimeters have a ratio of 2 : 1.   (4) Their corresponding angles have a ratio of 2 : 1.

19. In the diagram below of $\triangle PRT$, Q is a point on $\overline{PR}$, S is a point on $\overline{TR}$, $\overline{QS}$ is drawn, and $\angle RPT \cong \angle RSQ$.

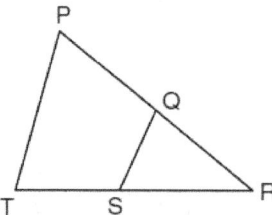

Which reason justifies the conclusion that $\triangle PRT \sim \triangle SRQ$?
(1)   AA   (2)   ASA   (3)   SAS   (4)   SSS

20. In the diagram of $\triangle ABC$ and $\triangle EDC$ below, $\overline{AE}$ and $\overline{BD}$ intersect at C, and $\angle CAB \cong \angle CED$.

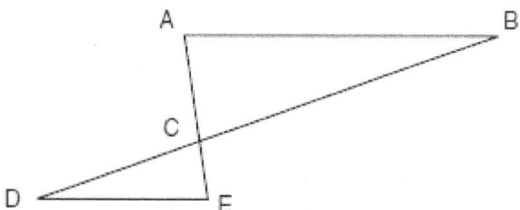

Which method can be used to show that $\triangle ABC$ must be similar to $\triangle EDC$?
(1)   SAS   (2)   AA   (3)   SSS   (4)   HL

21. In $\triangle ABC$, point D is on $\overline{AB}$, and point E is on $\overline{BC}$ such that $\overline{DE} \parallel \overline{AC}$. If $DB = 2$, $DA = 7$, and $DE = 3$, what is the length of $\overline{AC}$?
(1)   8   (2)   9   (3)   10.5   (4)   13.5

22. In the diagram below of $\triangle ACT$, D is the midpoint of $\overline{AC}$, O is the midpoint of $\overline{AT}$, and G is the midpoint of $\overline{CT}$.

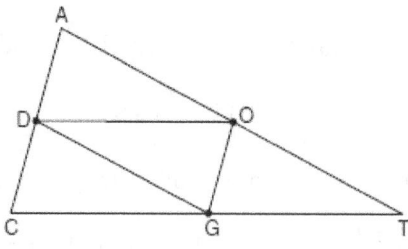

If $AC = 10$, $AT = 18$, and $CT = 22$, what is the perimeter of parallelogram $CDOG$?
(1)   21   (2)   25   (3)   32   (4)   40

23. In the diagram below, $\overline{SQ}$ and $\overline{PR}$ intersect at T, $\overline{PQ}$ is drawn, and $\overline{PS} \parallel \overline{QR}$.

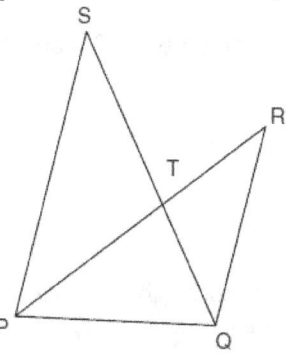

What technique can be used to prove that $\triangle PST \sim \triangle RQT$?
(1)   SAS   (2)   SSS   (3)   ASA   (4)   AA

24. Given $\triangle ABC \sim \triangle DEF$ such that $\dfrac{AB}{DE} = \dfrac{3}{2}$. Which statement is *not* true?

(1) $\dfrac{BC}{EF} = \dfrac{3}{2}$ (2) $\dfrac{m\angle A}{m\angle D} = \dfrac{3}{2}$ (3) $\dfrac{\text{area of } \triangle ABC}{\text{area of } \triangle DEF} = \dfrac{9}{4}$ (4) $\dfrac{\text{perimeter of } \triangle ABC}{\text{perimeter of } \triangle DEF} = \dfrac{3}{2}$

25. In the diagram below, the length of the legs $\overline{AC}$ and $\overline{BC}$ of right triangle ABC are 6 cm and 8 cm, respectively. Altitude $\overline{CD}$ is drawn to the hypotenuse of $\triangle ABC$.

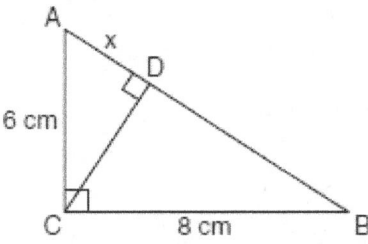

What is the length of $\overline{AD}$ to the *nearest tenth of a centimeter*?
(1)   3.6   (2)   6.0   (3)   6.4   (4)   4.0

26. In the diagram below of right triangle ACB, altitude $\overline{CD}$ is drawn to hypotenuse $\overline{AB}$.

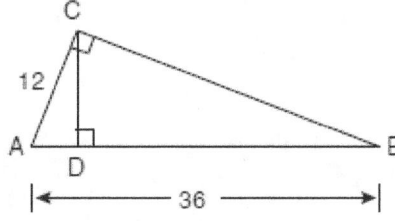

If $AB = 36$ and $AC = 12$, what is the length of $\overline{AD}$?
(1)   32   (2)   6   (3)   3   (4)   4

Show Work:

1. The degree measures of the angles of $\triangle ABC$ are represented by x, $3x$, and $5x - 54$. Find the value of x.

2. In the diagram below of $\triangle ABC$ with side $\overline{AC}$ extended through D, m$\angle A = 37$ and m$\angle BCD = 117$. Which side of $\triangle ABC$ is the longest side? Justify your answer.

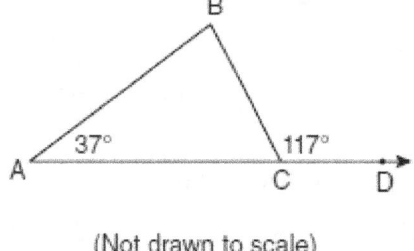

(Not drawn to scale)

3. In the diagram below of $\triangle TEM$, medians $\overline{TB}$, $\overline{EC}$, and $\overline{MA}$ intersect at D, and $TB = 9$. Find the length of $\overline{TD}$.

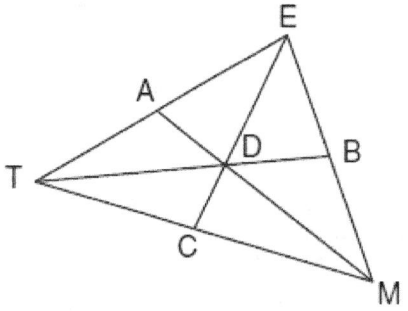

4. In $\triangle RST$, m$\angle RST = 46$ and $\overline{RS} \cong \overline{ST}$. Find m$\angle STR$.

5. Given: $\triangle ABC$ and $\triangle EDC$, C is the midpoint of $\overline{BD}$ and $\overline{AE}$
Prove: $\overline{AB} \parallel \overline{DE}$

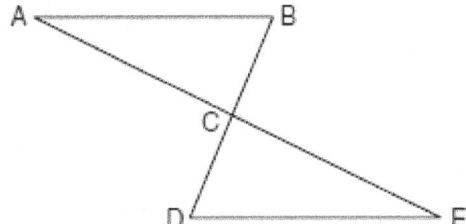

6. In the diagram of $\triangle ABC$ below, $AB = 10$, $BC = 14$, and $AC = 16$. Find the perimeter of the triangle formed by connecting the midpoints of the sides of $\triangle ABC$.

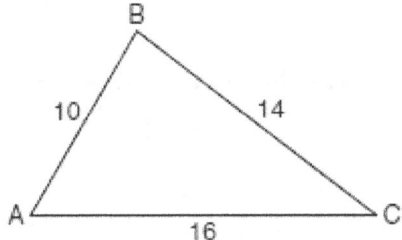

7. In the diagram below, $\triangle ABC \sim \triangle EFG$, $m\angle C = 4x + 30$, and $m\angle G = 5x + 10$. Determine the value of x.

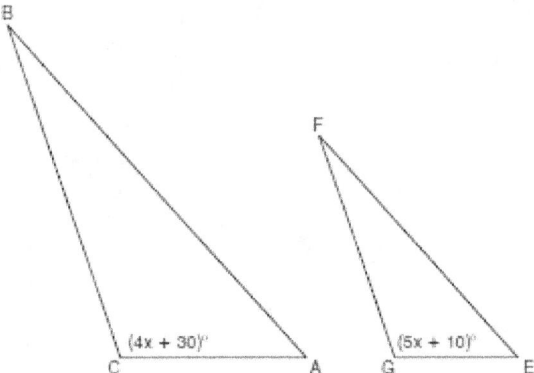

8. In the diagram below of $\triangle ACD$, E is a point on $\overline{AD}$ and B is a point on $\overline{AC}$, such that $\overline{EB} \parallel \overline{DC}$. If $AE = 3$, $ED = 6$, and $DC = 15$, find the length of $\overline{EB}$.

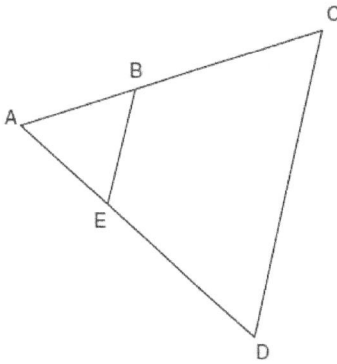

9. In the diagram below of right triangle ACB, altitude $\overline{CD}$ intersects $\overline{AB}$ at D. If $AD = 3$ and $DB = 4$, find the length of $\overline{CD}$ in simplest radical form.

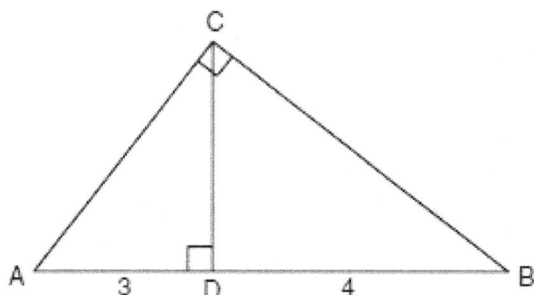

VI. POLYGONS

1. In the diagram below of parallelogram $ABCD$ with diagonals $\overline{AC}$ and $\overline{BD}$, $m\angle 1 = 45$ and $m\angle DCB = 120$.

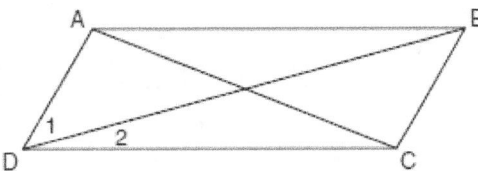

What is the measure of $\angle 2$?
(1)   15°   (2)   30°   (3)   45°   (4)   60°

2. A quadrilateral whose diagonals bisect each other and are perpendicular is a
(1) rhombus   (3) trapezoid
(2) rectangle   (4) Parallelogram

3. In the diagram below of parallelogram $STUV$, $SV = x + 3$, $VU = 2x - 1$, and $TU = 4x - 3$.

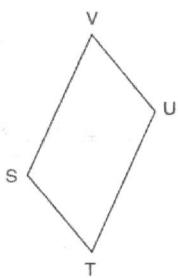

What is the length of $\overline{SV}$?
(1)   5   (2)   2   (3)   7   (4)   4

4. Isosceles trapezoid $ABCD$ has diagonals $\overline{AC}$ and $\overline{BD}$. If $AC = 5x + 13$ and $BD = 11x - 5$, what is the value of x?

(1)   28   (2) $10\dfrac{3}{4}$ (3)   3   (4) $\dfrac{1}{2}$

5. In the diagram below of trapezoid $RSUT$, $\overline{RS} \parallel \overline{TU}$, X is the midpoint of $\overline{RT}$, and V is the midpoint of $\overline{SU}$.

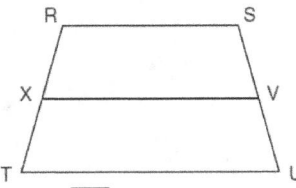

If $RS = 30$ and $XV = 44$, what is the length of $\overline{TU}$?
(1)   37   (2)   58   (3)   74   (4)   118

6. What is the measure of an interior angle of a regular octagon?
(1)   45°   (2)   60°   (3)   120°   (4)   135°

7. The pentagon in the diagram below is formed by five rays.

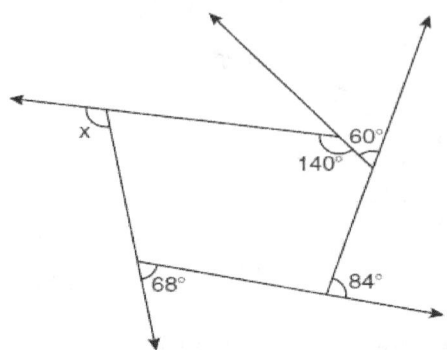

What is the degree measure of angle x?
(1)   72   (2)   96   (3)   108   (4)   112

Show Work:

1. Given: Quadrilateral $ABCD$, diagonal $\overline{AFEC}$, $\overline{AE} \cong \overline{FC}$, $\overline{BF} \perp \overline{AC}$, $\overline{DE} \perp \overline{AC}$, $\angle 1 \cong \angle 2$
Prove: $ABCD$ is a parallelogram.

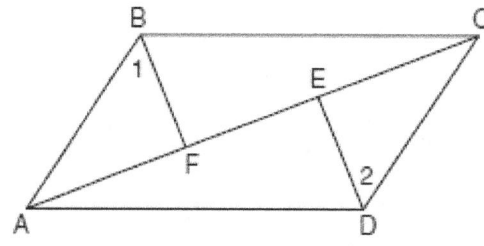

2. Given: $JKLM$ is a parallelogram.
   $\overline{JM} \cong \overline{LN}$
   $\angle LMN \cong \angle LNM$

Prove: $JKLM$ is a rhombus.

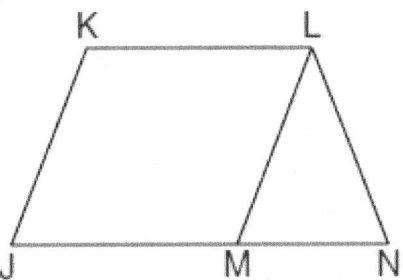

3. In the diagram below of isosceles trapezoid $DEFG$, $\overline{DE} \parallel \overline{GF}$, $DE = 4x - 2$, $EF = 3x + 2$, $FG = 5x - 3$, and $GD = 2x + 5$. Find the value of x.

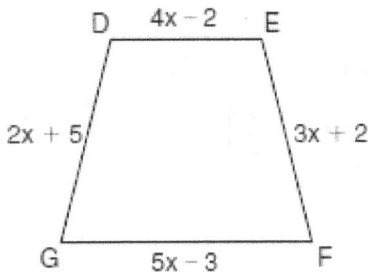

4. In the diagram below, quadrilateral $STAR$ is a rhombus with diagonals $\overline{SA}$ and $\overline{TR}$ intersecting at E. $ST = 3x + 30$, $SR = 8x - 5$, $SE = 3z$, $TE = 5z + 5$, $AE = 4z - 8$, m$\angle RTA = 5y - 2$, and m$\angle TAS = 9y + 8$. Find SR, RT, and m$\angle TAS$.

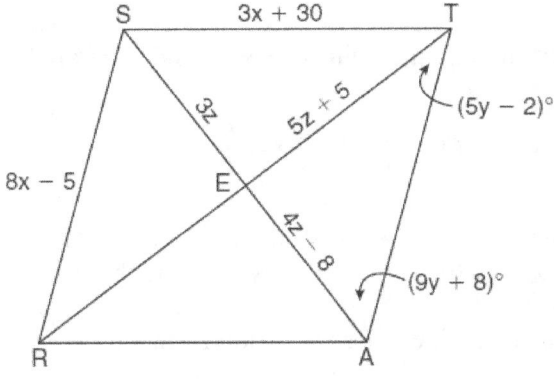

VII. COORDINATE GEOMETRY

1. Line segment AB has endpoints $A(2,-3)$ and $B(-4,6)$. What are the coordinates of the midpoint of $\overline{AB}$?

(1) $(-2,3)$ (2) $\left(-1, 1\frac{1}{2}\right)$ (3) $(-1,3)$ (4) $\left(3, 4\frac{1}{2}\right)$

2. Square $LMNO$ is shown in the diagram below.

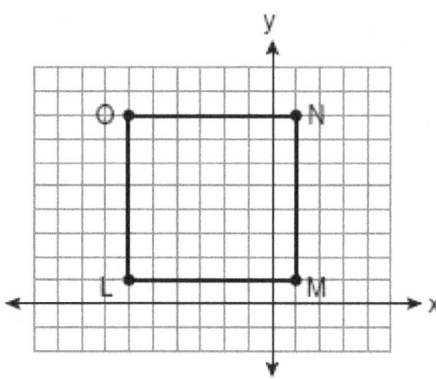

What are the coordinates of the midpoint of diagonal $\overline{LN}$?

(1) $\left(4\frac{1}{2}, -2\frac{1}{2}\right)$ (2) $\left(-3\frac{1}{2}, 3\frac{1}{2}\right)$ (3) $\left(-2\frac{1}{2}, 3\frac{1}{2}\right)$ (4) $\left(-2\frac{1}{2}, 4\frac{1}{2}\right)$

3. The endpoints of $\overline{CD}$ are $C(-2,-4)$ and $D(6,2)$. What are the coordinates of the midpoint of $\overline{CD}$?
(1) $(2,3)$ (2) $(2,-1)$ (3) $(4,-2)$ (4) $(4,3)$

4. If the endpoints of $\overline{AB}$ are $A(-4,5)$ and $B(2,-5)$, what is the length of $\overline{AB}$?
(1) $2\sqrt{34}$ (2) 2 (3) $\sqrt{61}$ (4) 8

5. What is the distance between the points $(-3,2)$ and $(1,0)$?
(1) $2\sqrt{2}$ (2) $2\sqrt{3}$ (3) $5\sqrt{2}$ (4) $2\sqrt{5}$

6. What is the equation of a line that passes through the point $(-3,-11)$ and is parallel to the line whose equation is $2x-y=4$?

(1) $y=2x+5$ (2) $y=2x-5$ (3) $y=\frac{1}{2}x+\frac{25}{2}$ (4) $y=-\frac{1}{2}x-\frac{25}{2}$

7. The lines $3y+1=6x+4$ and $2y+1=x-9$ are
(1)   parallel (2)   perpendicular (3)   the same line   (4)   neither parallel nor perpendicular

8. What is the slope of a line perpendicular to the line whose equation is $5x+3y=8$?
(1) $\frac{5}{3}$ (2) $\frac{3}{5}$ (3) $-\frac{3}{5}$ (4) $-\frac{5}{3}$

9. What is an equation of the line that passes through the point $(-2, 5)$ and is perpendicular to the line whose equation is $y = \frac{1}{2}x + 5$?

(1) $y = 2x + 1$ (2) $y = -2x + 1$ (3) $y = 2x + 9$ (4) $y = -2x - 9$

10. Which equation represents a line perpendicular to the line whose equation is $2x + 3y = 12$?
(1) $6y = -4x + 12$ (2) $2y = 3x + 6$ (3) $2y = -3x + 6$ (4) $3y = -2x + 12$

11. What is the equation of a line that is parallel to the line whose equation is $y = x + 2$?
(1) $x + y = 5$ (2) $2x + y = -2$ (3) $y - x = -1$ (4) $y - 2x = 3$

12. What is the slope of a line perpendicular to the line whose equation is $y = -\frac{2}{3}x - 5$?

(1) $-\frac{3}{2}$ (2) $-\frac{2}{3}$ (3) $\frac{2}{3}$ (4) $\frac{3}{2}$

13. Which equation represents a line parallel to the line whose equation is $2y - 5x = 10$?
(1) $5y - 2x = 25$ (2) $5y + 2x = 10$ (3) $4y - 10x = 12$ (4) $2y + 10x = 8$

14. What is an equation of the line that contains the point $(3, -1)$ and is perpendicular to the line whose equation is $y = -3x + 2$?

(1) $y = -3x + 8$ (2) $y = -3x$ (3) $y = \frac{1}{3}x$ (4) $y = \frac{1}{3}x - 2$

15. What is the slope of a line that is perpendicular to the line whose equation is $3x + 4y = 12$?

(1) $\frac{3}{4}$ (2) $-\frac{3}{4}$ (3) $\frac{4}{3}$ (4) $-\frac{4}{3}$

16. Two lines are represented by the equations $-\frac{1}{2}y = 6x + 10$ and $y = mx$. For which value of m will the lines be parallel?
(1)   -12   (2)   -3   (3)   3   (4)   12

17. The vertices of $\triangle ABC$ are $A(-1, -2)$, $B(-1, 2)$ and $C(6, 0)$. Which conclusion can be made about the angles of $\triangle ABC$?
(1) $m\angle A = m\angle B$ (2) $m\angle A = m\angle C$ (3) $m\angle ACB = 90$ (4) $m\angle ABC = 60$

Show Work:

1. The endpoints of $\overline{PQ}$ are $P(-3, 1)$ and $Q(4, 25)$. Find the length of $\overline{PQ}$.

2. Triangle ABC has coordinates $A(-6, 2)$, $B(-3, 6)$, and $C(5, 0)$. Find the perimeter of the triangle. Express your answer in simplest radical form. [The use of the grid below is optional.]

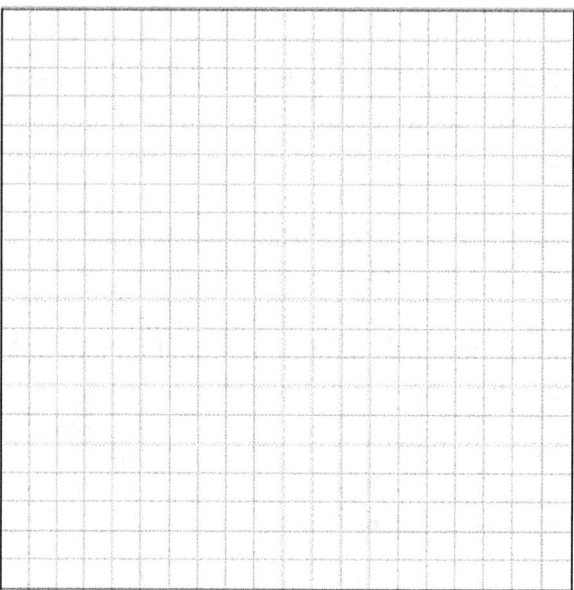

3. In the diagram below of circle C, $\overline{QR}$ is a diameter, and $Q(1, 8)$ and $C(3.5, 2)$ are points on a coordinate plane. Find and state the coordinates of point R.

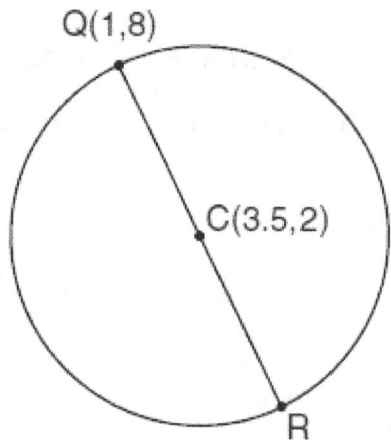

4. Find an equation of the line passing through the point $(5, 4)$ and parallel to the line whose equation is $2x + y = 3$.

5. Write an equation of the line that passes through the point $(6, -5)$ and is parallel to the line whose equation is $2x - 3y = 11$.

6. On the set of axes below, graph and label $\triangle DEF$ with vertices at $D(-4, -4)$, $E(-2, 2)$, and $F(8, -2)$. If G is the midpoint of $\overline{EF}$ and H is the midpoint of $\overline{DF}$, state the coordinates of G and H and label each point on your graph. Explain why $\overline{GH} \parallel \overline{DE}$.

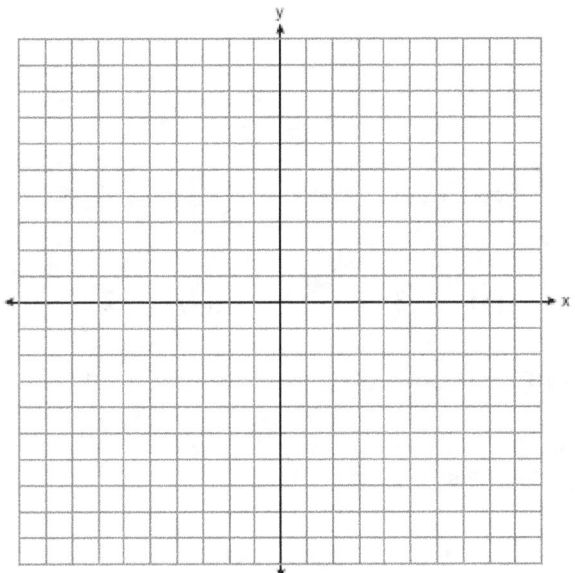

VIII. CIRCLE

1. In the diagram below of circle O, chords $\overline{AD}$ and $\overline{BC}$ intersect at E, $m\overset{\frown}{AC} = 87$, and $m\overset{\frown}{BD} = 35$.

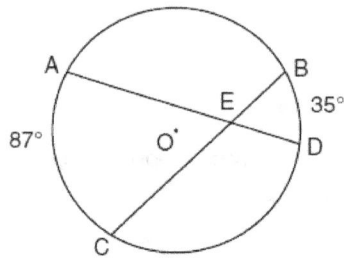

What is the degree measure of $\angle CEA$?
(1)   87   (2)   61   (3)   43.5   (4)   26

2. In the diagram below, $\overline{PS}$ is a tangent to circle O at point S, $\overline{PQR}$ is a secant, $PS = x$, $PQ = 3$, and $PR = x + 18$.

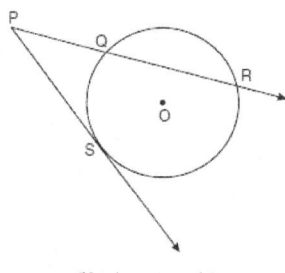

(Not drawn to scale)

What is the length of $\overline{PS}$?
(1)   6   (2)   9   (3)   3   (4)   27

3. In the diagram below, tangent $\overline{AB}$ and secant $\overline{ACD}$ are drawn to circle O from an external point A, $AB = 8$, and $AC = 4$.

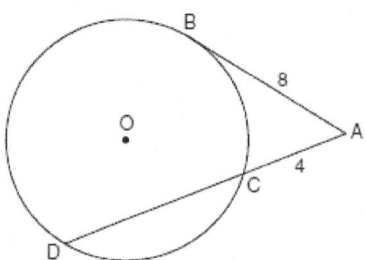

What is the length of $\overline{CD}$?
(1)   16   (2)   13   (3)   12   (4)   10

4. In the diagram of circle O below, chord $\overline{AB}$ intersects chord $\overline{CD}$ at E, $DE = 2x + 8$, $EC = 3$, $AE = 4x - 3$, and $EB = 4$.

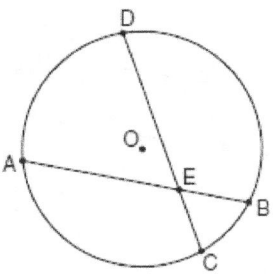

What is the value of x?
(1)   1   (2)   3.6   (3)   5   (4)   10.25

5. In the diagram below, tangent $\overline{PA}$ and secant $\overline{PBC}$ are drawn to circle O from external point P.

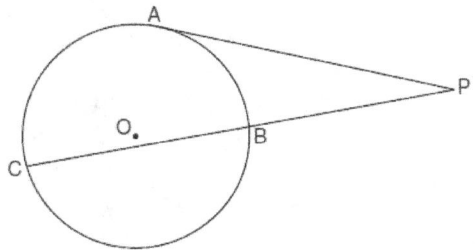

If $PB = 4$ and $BC = 5$, what is the length of $\overline{PA}$?
(1)   20   (2)   9   (3)   8   (4)   6

6. In the diagram below, circle O has a radius of 5, and $CE = 2$. Diameter $\overline{AC}$ is perpendicular to chord $\overline{BD}$ at E.

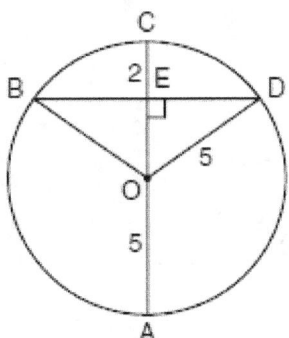

What is the length of $\overline{BD}$?
(1)   12   (2)   10   (3)   8   (4)   4

7. In the diagram of circle O below, chords $\overline{AB}$ and $\overline{CD}$ are parallel, and $\overline{BD}$ is a diameter of the circle.

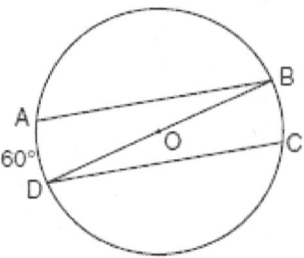

If $m\overset{\frown}{AD} = 60$, what is $m\angle CDB$?
(1)   20   (2)   30   (3)   60   (4)   120

8. In the diagram of circle O below, chord $\overline{CD}$ is parallel to diameter $\overline{AOB}$ and $m\overset{\frown}{AC} = 30$.

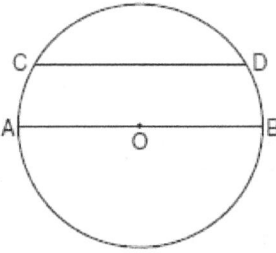

What is $m\overset{\frown}{CD}$?
(1)   150   (2)   120   (3)   100   (4)   60

9. In the diagram below, $\triangle ABC$ is inscribed in circle P. The distances from the center of circle P to each side of the triangle are shown.

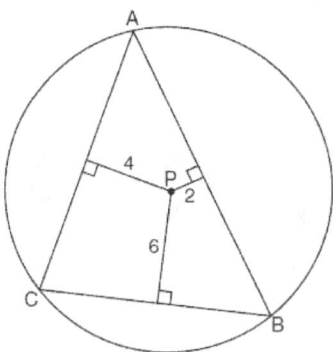

Which statement about the sides of the triangle is true?
(1) $AB > AC > BC$ (2) $AB < AC$ and $AC > BC$ (3) $AC > AB > BC$ (4) $AC = AB$ and $AB > BC$

10. How many common tangent lines can be drawn to the two externally tangent circles shown below?

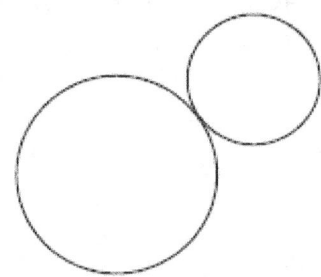

(1)   1   (2)   2   (3)   3   (4)   4

Show Work:

1. In the diagram below of circle O, chords $\overline{DF}$, $\overline{DE}$, $\overline{FG}$, and $\overline{EG}$ are drawn such that $m\overarc{DF} : m\overarc{FE} : m\overarc{EG} : m\overarc{GD} = 5:2:1:7$. Identify one pair of inscribed angles that are congruent to each other and give their measure.

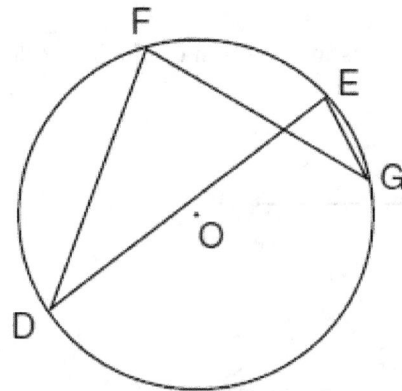

2. In the diagram below, circles X and Y have two tangents drawn to them from external point T. The points of tangency are C, A, S, and E. The ratio of TA to AC is $1:3$. If $TS = 24$, find the length of $\overline{SE}$.

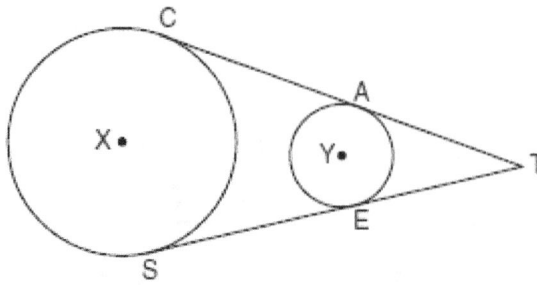

(Not drawn to scale)

3. In the diagram below, quadrilateral $ABCD$ is inscribed in circle O, $\overline{AB} \parallel \overline{DC}$, and diagonals $\overline{AC}$ and $\overline{BD}$ are drawn. Prove that $\triangle ACD \cong \triangle BDC$.

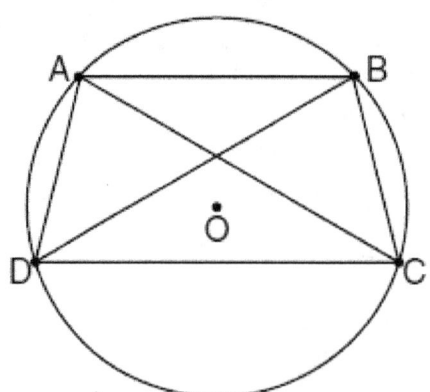

IX. CONSTRUCTIONS AND LOCI

1. The diagram below shows the construction of the perpendicular bisector of $\overline{AB}$.

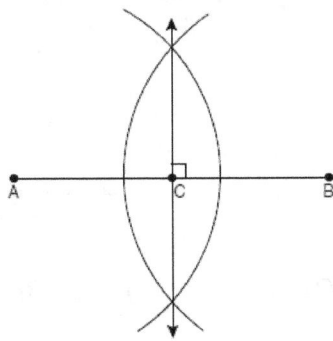

Which statement is *not* true?

(1) $AC = CB$ (2) $CB = \frac{1}{2} AB$ (3) $AC = 2AB$ (4) $AC + CB = AB$

2. Which illustration shows the correct construction of an angle bisector?

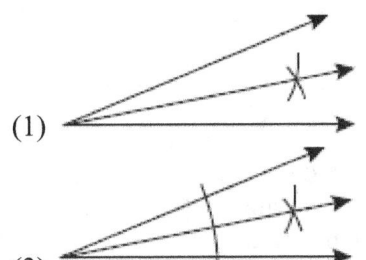

(1)

(2)

(3)

(4)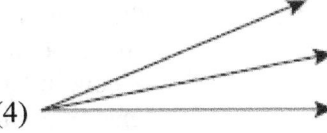

3. The diagram below shows the construction of the bisector of $\angle ABC$.

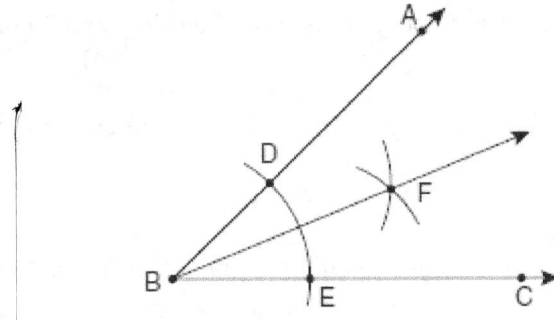

Which statement is *not* true?

(1) $m\angle EBF = \frac{1}{2} m\angle ABC$ (2) $m\angle DBF = \frac{1}{2} m\angle ABC$ (3) $m\angle EBF = m\angle ABC$ (4) $m\angle DBF = m\angle EBF$

4. Based on the construction below, which statement must be true?

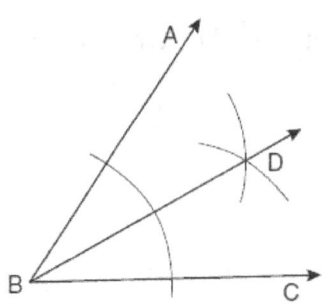

(1) $m\angle ABD = \frac{1}{2}m\angle CBD$ (2) $m\angle ABD = m\angle CBD$ (3) $m\angle ABD = m\angle ABC$ (4) $m\angle CBD = \frac{1}{2}m\angle ABD$

5. Which geometric principle is used to justify the construction below?

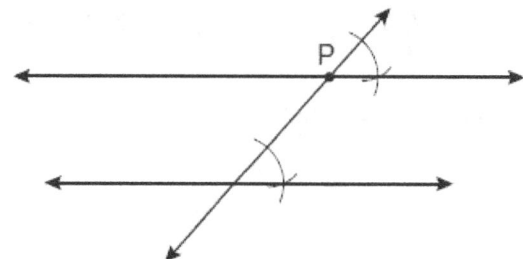

(1) A line perpendicular to one of two parallel lines is perpendicular to the other.

(2) Two lines are perpendicular if they intersect to form congruent adjacent angles.

(3) When two lines are intersected by a transversal and alternate interior angles are congruent, the lines are parallel.

(4) When two lines are intersected by a transversal and the corresponding angles are congruent, the lines are parallel.

6. In a coordinate plane, how many points are both 5 units from the origin and 2 units from the x-axis?
(1)   1   (2)   2   (3)   3   (4)   4

7. Towns A and B are 16 miles apart. How many points are 10 miles from town A and 12 miles from town B?
(1)   1   (2)   2   (3)   3   (4)   0

8. What are the center and radius of a circle whose equation is $(x - A)^2 + (y - B)^2 = C$?

(1)   center $= (A, B)$; radius $= C$ (3)   center $= (A, B)$; radius $= \sqrt{C}$

(2)   center $= (-A, -B)$; radius $= C$ (4)   center $= (-A, -B)$; radius $= \sqrt{C}$

9. The diameter of a circle has endpoints at $(-2, 3)$ and $(6, 3)$. What is an equation of the circle?
(1) $(x - 2)^2 + (y - 3)^2 = 16$ (3) $(x + 2)^2 + (y + 3)^2 = 16$
(2) $(x - 2)^2 + (y - 3)^2 = 4$ (4) $(x + 2)^2 + (y + 3)^2 = 4$

10. What is an equation of a circle with its center at $(-3, 5)$ and a radius of 4?

   (1) $(x-3)^2 + (y+5)^2 = 16$ (3) $(x-3)^2 + (y+5)^2 = 4$

   (2) $(x+3)^2 + (y-5)^2 = 16$ (4) $(x+3)^2 + (y-5)^2 = 4$

11. A circle is represented by the equation $x^2 + (y+3)^2 = 13$. What are the coordinates of the center of the circle and the length of the radius?

   (1) $(0, 3)$ and 13 (2) $(0, 3)$ and $\sqrt{13}$ (3) $(0, -3)$ and 13 (4) $(0, -3)$ and $\sqrt{13}$

12. What are the center and the radius of the circle whose equation is $(x-3)^2 + (y+3)^2 = 36$

   (1)   center $= (3, -3)$; radius $= 6$ (3)   center $= (3, -3)$; radius $= 36$

   (2)   center $= (-3, 3)$; radius $= 6$ (4)   center $= (-3, 3)$; radius $= 36$

13. Which equation represents circle K shown in the graph below?

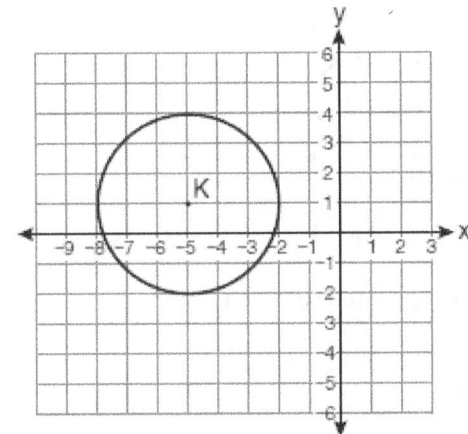

   (1) $(x+5)^2 + (y-1)^2 = 3$ (3) $(x-5)^2 + (y+1)^2 = 3$

   (2) $(x+5)^2 + (y-1)^2 = 9$ (4) $(x-5)^2 + (y+1)^2 = 9$

14. Which equation represents the circle whose center is $(-2, 3)$ and whose radius is 5?

   (1) $(x-2)^2 + (y+3)^2 = 5$ (3) $(x+2)^2 + (y-3)^2 = 25$

   (2) $(x+2)^2 + (y-3)^2 = 5$ (4) $(x-2)^2 + (y+3)^2 = 25$

15. Given the system of equations:

$$y = x^2 - 4x$$

$$x = 4$$

The number of points of intersection is

   (1)   1   (2)   2   (3)   3   (4)   0

16. Given the equations: $y = x^2 - 6x + 10$

$$y + x = 4$$

What is the solution to the given system of equations?

   (1) $(2, 3)$ (2) $(3, 2)$ (3) $(2, 2)$ and $(1, 3)$ (4) $(2, 2)$ and $(3, 1)$

17. The equation of a circle is $(x-2)^2 + (y+4)^2 = 4$. Which diagram is the graph of the circle?

(1)

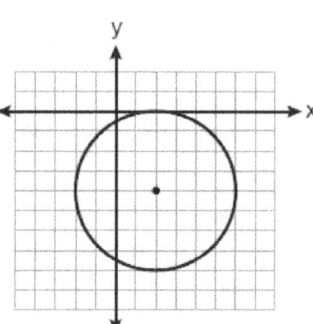

(3)

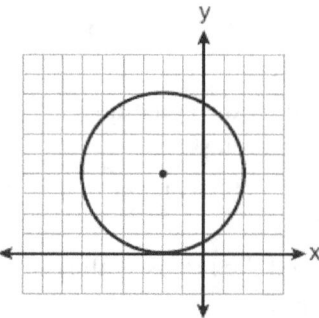

(2)

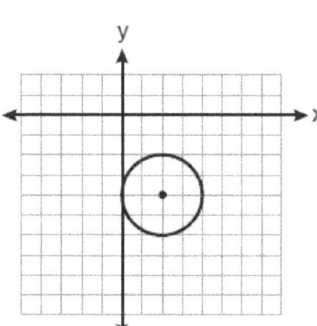

(4)

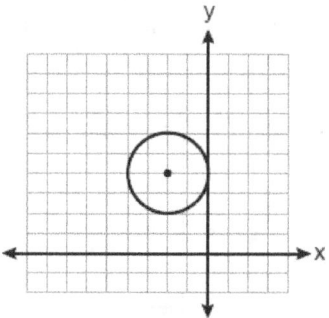

18. Which graph could be used to find the solution to the following system of equations?

$$y = -x + 2$$

$$y = x^2$$

(1)

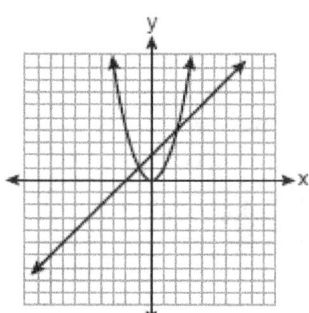

(3)

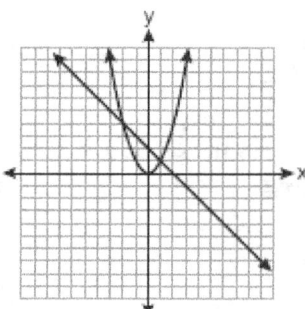

(2)

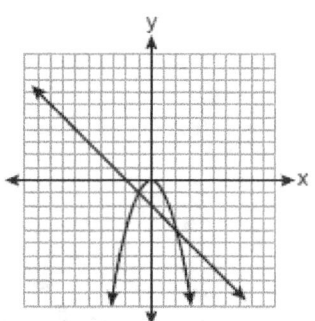

(4)

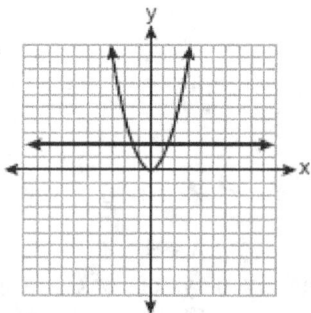

Show Work:

1. Using a compass and straightedge, construct the bisector of the angle shown below. [*Leave all construction marks*.]

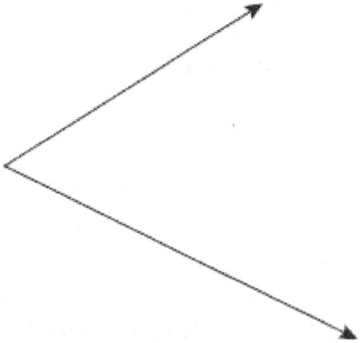

2. Using a compass and straightedge, construct a line that passes through point *P* and is perpendicular to line *m*. [Leave all construction marks.]

3. Using a compass and straightedge, and $\overline{AB}$ below, construct an equilateral triangle with all sides congruent to $\overline{AB}$. [Leave all construction marks.]

4. The length of $\overline{AB}$ is 3 inches. On the diagram below, sketch the points that are equidistant from A and B and sketch the points that are 2 inches from A. Label with an **X** all points that satisfy both conditions.

A ●————————————————————● B

5. A city is planning to build a new park. The park must be equidistant from school A at $(3, 3)$ and school B at $(3, -5)$. The park also must be exactly 5 miles from the center of town, which is located at the origin on the coordinate graph. Each unit on the graph represents 1 mile. On the set of axes below, sketch the compound loci and label with an **X** all possible locations for the new park.

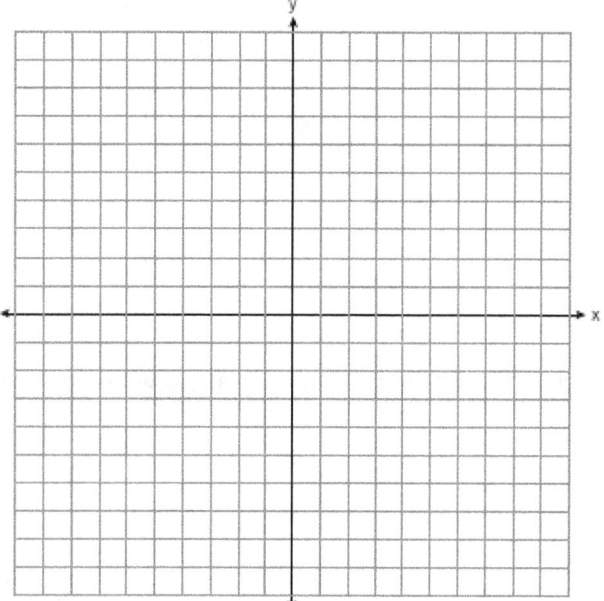

6. On the grid below, graph the points that are equidistant from both the x and y axes and the points that are 5 units from the origin. Label with an **X** all points that satisfy both conditions.

7. Write an equation of the perpendicular bisector of the line segment whose endpoints are $(-1, 1)$ and $(7, -5)$. [The use of the grid below is optional]

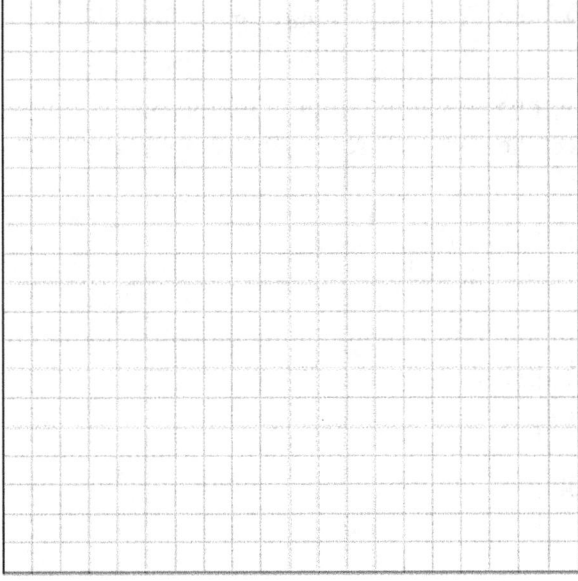

X. TRANSFORMATION

1. Triangle ABC has vertices $A(1,3)$, $B(0,1)$, and $C(4,0)$. Under a translation, A', the image point of A, is located at $(4,4)$. Under this same translation, point C' is located at

(1) $(7,1)$ (2) $(5,3)$ (3) $(3,2)$ (4) $(1,-1)$

2. A polygon is transformed according to the rule: $(x,y) \rightarrow (x+2,y)$. Every point of the polygon moves two units in which direction?

(1)   up   (2)   down   (3)   left   (4)   right

3. In the diagram below, under which transformation will $\triangle A'B'C'$ be the image of $\triangle ABC$?

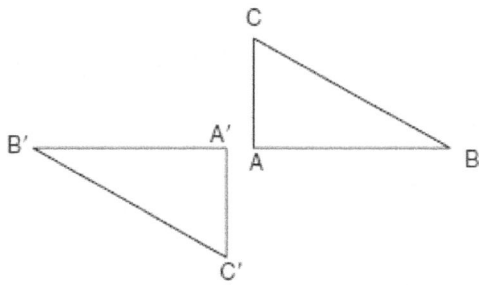

(1)   rotation   (2)   dilation   (3)   translation   (4)   glide reflection

4. Point A is located at $(4,-7)$. The point is reflected in the x-axis. Its image is located at

(1) $(-4,7)$ (2) $(-4,-7)$ (3) $(4,7)$ (4) $(7,-4)$

5. Which transformation produces a figure similar but not congruent to the original figure?

(1) $T_{1,3}$ (2) $D_{\frac{1}{2}}$ (3) $R_{90°}$ (4) $r_{y=x}$

6. On the set of axes below, Geoff drew rectangle $ABCD$. He will transform the rectangle by using the translation $(x,y) \rightarrow (x+2,y+1)$ and then will reflect the translated rectangle over the x-axis.

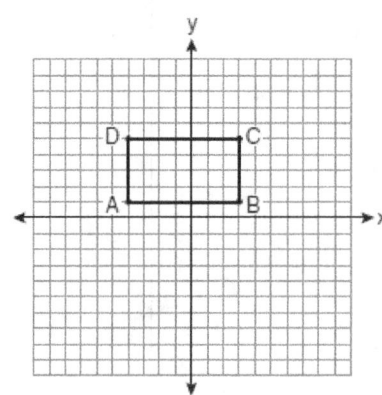

What will be the area of the rectangle after these transformations?

(1) exactly 28 square units   (3) greater than 28 square units
(2) less than 28 square units   (4) It cannot be determined from the information given.

7. Which expression best describes the transformation shown in the diagram below?

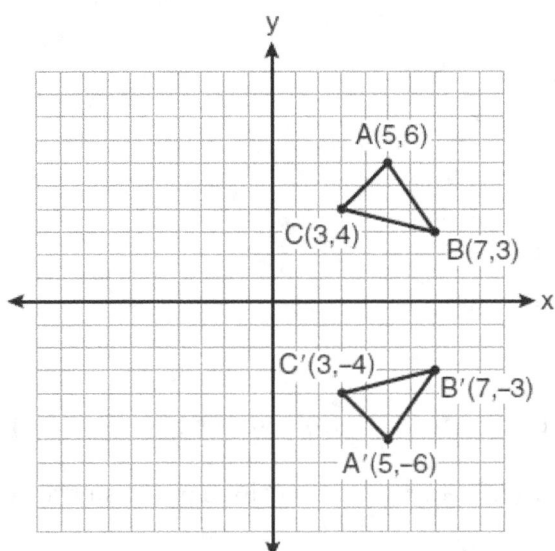

(1) same orientation; reflection   (3) same orientation; translation
(2) opposite orientation; reflection   (4) opposite orientation; translation

8. The endpoints of $\overline{AB}$ are $A(3, 2)$ and $B(7, 1)$. If $\overline{A''B''}$ is the result of the transformation of $\overline{AB}$ under $D_2 \circ T_{-4,3}$ what are the coordinates of A'' and B''?

(1) $A''(-2, 10)$ and $B''(6, 8)$ (3) $A''(2, 7)$ and $B''(10, 5)$
(2) $A''(-1, 5)$ and $B''(3, 4)$ (4) $A''(14, -2)$ and $B''(22, -4)$

9. After a composition of transformations, the coordinates $A(4, 2)$, $B(4, 6)$, and $C(2, 6)$ become $A''(-2, -1)$, $B''(-2, -3)$, and $C''(-1, -3)$, as shown on the set of axes below.

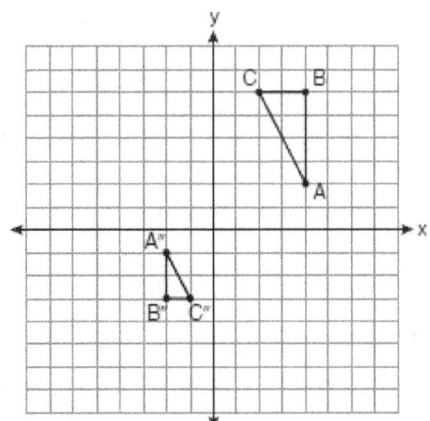

Which composition of transformations was used?
(1) $R_{180°} \circ D_2$ (2) $R_{90°} \circ D_2$ (3) $D_{\frac{1}{2}} \circ R_{180°}$ (4) $D_{\frac{1}{2}} \circ R_{90°}$

10. In the diagram below, which transformation was used to map $\triangle ABC$ to $\triangle A'B'C'$?

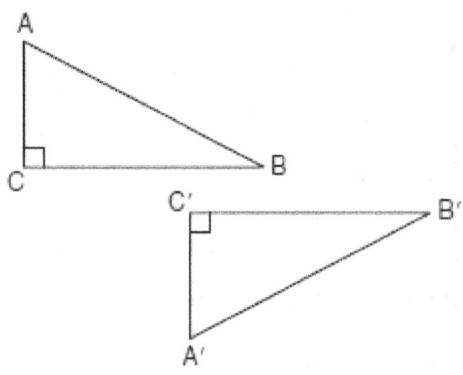

(1) dilation   (2) rotation   (3) reflection   (4) glide reflection

11. What is the image of point $A(4,2)$ after the composition of transformations defined by $R_{90°} \circ r_{y=x}$?

(1) $(-4,2)$ (2) $(4,-2)$ (3) $(-4,-2)$ (4) $(2,-4)$

12. Which transformation is *not* always an isometry?

(1)   rotation   (2)   dilation   (3)   reflection   (4)   translation

Show Work:

1. Triangle DEG has the coordinates $D(1,1)$, $E(5,1)$, and $G(5,4)$. Triangle DEG is rotated 90° about the origin to form $\triangle D'E'G'$. On the grid below, graph and label $\triangle DEG$ and $\triangle D'E'G'$. State the coordinates of the vertices D', E', and G'. Justify that this transformation preserves distance.

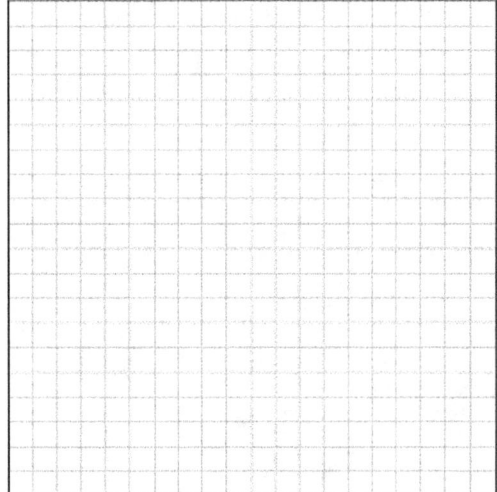

2. In $\triangle KLM$, $m\angle K = 36$ and $KM = 5$. The transformation D_2 is performed on $\triangle KLM$ to form $\triangle K'L'M'$. Find $m\angle K'$. Justify your answer. Find the length of $\overline{K'M'}$. Justify your answer.

3. The coordinates of the vertices of parallelogram $ABCD$ are $A(-2, 2)$, $B(3, 5)$, $C(4, 2)$, and $D(-1, -1)$. State the coordinates of the vertices of parallelogram $A''B''C''D''$ that result from the transformation $r_{y-axis} \circ T_{2,-3}$. [The use of the set of axes below is optional.]

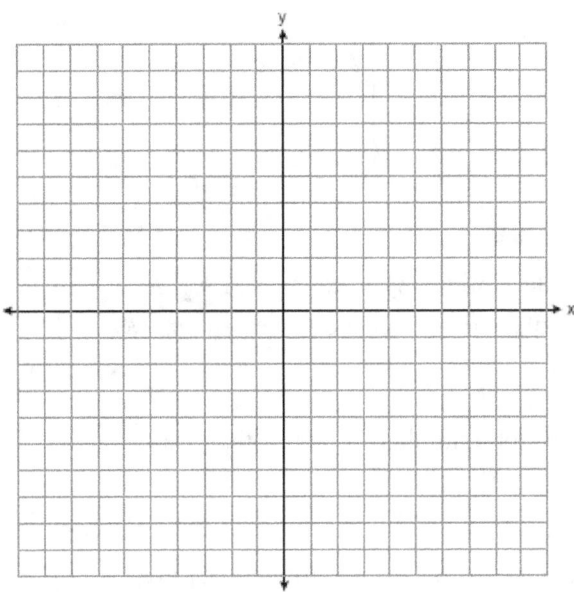

4. On the set of axes below, solve the following system of equations graphically for all values of x and y.

$$y = (x - 2)^2 + 4$$

$$4x + 2y = 14$$

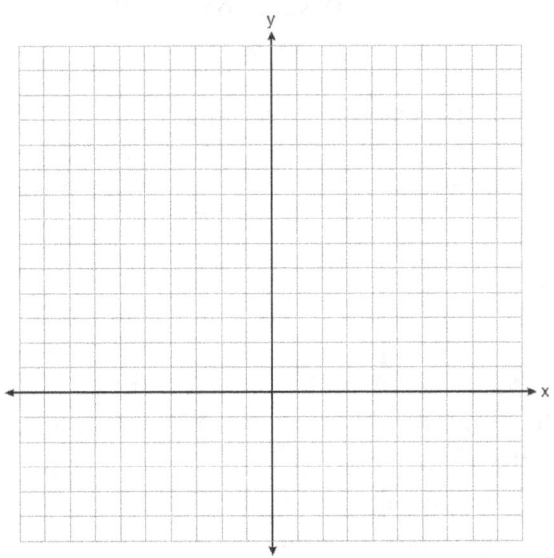

XI. SOLID GEOMETRY

1. Lines k_1 and k_2 intersect at point E. Line m is perpendicular to lines k_1 and k_2 at point E.

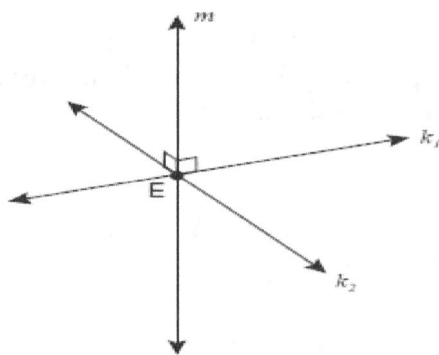

Which statement is always true?
(1) Lines k_1 and k_2 are perpendicular.
(2) Line m is parallel to the plane determined by lines k_1 and k_2.
(3) Line m is perpendicular to the plane determined by lines k_1 and k_2.
(4) Line m is coplanar with lines k_1 and k_2.

2. Point P is on line m. What is the total number of planes that are perpendicular to line m and pass through point P?
(1)   1   (2)   2   (3)   0   (4) infinite

3. If two different lines are perpendicular to the same plane, they are
(1) collinear   (2) coplanar   (3) congruent   (4) consecutive

4. In plane $\mathcal{P}$, lines m and n intersect at point A. If line k is perpendicular to line m and line n at point A, then line k is
(1)   contained in plane $\mathcal{P}$ (3) perpendicular to plane $\mathcal{P}$
(2)   parallel to plane $\mathcal{P}$ (4) skew to plane $\mathcal{P}$

5. In the diagram below, line k is perpendicular to plane $\mathcal{P}$ at point T.

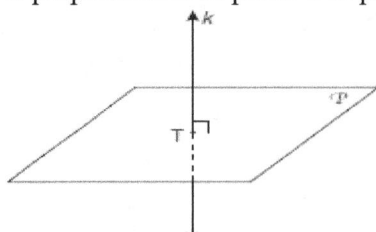

Which statement is true?
(1) Any point in plane $\mathcal{P}$ also will be on line k.
(2) Only one line in plane $\mathcal{P}$ will intersect line k.
(3) All planes that intersect plane $\mathcal{P}$ will pass through T.
(4) Any plane containing line k is perpendicular to plane $\mathcal{P}$.

6. Line k is drawn so that it is perpendicular to two distinct planes, P and R. What must be true about planes P and R?
(1) Planes P and R are skew.   (2) Planes P and R are parallel.
(3) Planes P and R are perpendicular.   (4) Plane P intersects plane R but is not perpendicular to plane R.

7. In three-dimensional space, two planes are parallel and a third plane intersects both of the parallel planes. The intersection of the planes is a
(1) plane   (2) point   (3) pair of parallel lines   (4) pair of intersecting lines

8. Through a given point, P, on a plane, how many lines can be drawn that are perpendicular to that plane?
(1)   1   (2)   2   (3)   more than 2   (4)   none

XIII. GEOMETIC MEASUREMENTS

1. The figure in the diagram below is a triangular prism.

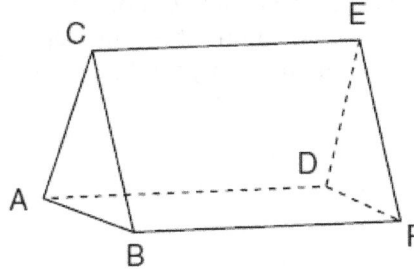

Which statement must be true?
 (1) $\overline{DE} \cong \overline{AB}$ (2) $\overline{AD} \cong \overline{BC}$ (3) $\overline{AD} \parallel \overline{CE}$ (4) $\overline{DE} \parallel \overline{BC}$

2. A right circular cylinder has a volume of 1,000 cubic inches and a height of 8 inches. What is the radius of the cylinder to the *nearest tenth of an inch*?
(1)   6.3   (2)   11.2   (3)   19.8   (4)   39.8

3. The lateral faces of a regular pyramid are composed of
(1)   squares   (2)   rectangles   (3)   congruent right triangles   (4)   congruent isosceles triangles

4. Which expression represents the volume, in cubic centimeters, of the cylinder represented in the diagram below?

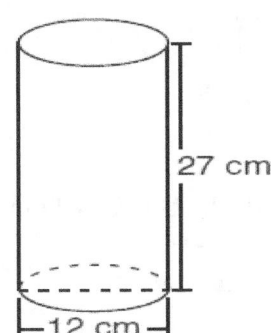

27 cm

12 cm

(1) 162π (2) 324π (3) 972π (4) $3,888\pi$

5. In the diagram below, a right circular cone has a diameter of 8 inches and a height of 12 inches.

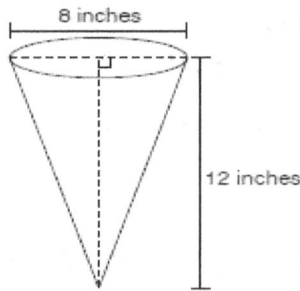

What is the volume of the cone to the *nearest cubic inch*?
(1)   201   (2)   481   (3)   603   (4)   804

Show Work:

1. Tim has a rectangular prism with a length of 10 centimeters, a width of 2 centimeters, and an unknown height. He needs to build another rectangular prism with a length of 5 centimeters and the same height as the original prism. The volume of the two prisms will be the same. Find the width, in centimeters, of the new prism.

2. The volume of a cylinder is 12,566.4 cm³. The height of the cylinder is 8 cm. Find the radius of the cylinder to the *nearest tenth of a centimeter*.

3. A regular pyramid with a square base is shown in the diagram below.

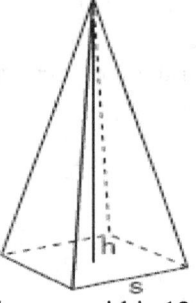

A side, *s*, of the base of the pyramid is 12 meters, and the height, *h*, is 42 meters. What is the volume of the pyramid in cubic meters?

I. LOGIC

1. What is the negation of the statement "The Sun is shining"?
(1) It is cloudy.   (3) It is not raining.
(2) It is daytime.   ***(4) The Sun is not shining.**

> Negation: not

2. What is the negation of the statement "Squares are parallelograms"?
(1) Parallelograms are squares.   ***(3) It is not the case that squares are parallelograms.**
(2) Parallelograms are not squares.   (4) It is not the case that parallelograms are squares.

> Negation: It is not the case that …

3. What is the contrapositive of the statement, "If I am tall, then I will bump my head"?
(1) If I bump my head, then I am tall.   (3) If I am tall, then I will not bump my head.
(2) If I do not bump my head, then I am tall.   ***(4) If I do not bump my head, then I am not tall.**

> Original: p → q , Contrapositive: not q → not p

4. What is the inverse of the statement "If two triangles are not similar, their corresponding angles are not congruent"?
(1) If two triangles are similar, their   ***(3) If two triangles are similar, their**
      corresponding angles are not congruent.   **corresponding angles are congruent.**
(2) If corresponding angles of two triangles are   (4) If corresponding angles of two triangles
      not congruent, the triangles are not similar.   are congruent, the triangles are similar.

> Original: p → q ,   Inverse: not p → not q ;
> Original: not p → not q ,   Inverse: p → q .

Show Work:

1. Given: Two is an even integer or three is an even integer.
Determine the truth value of this disjunction. Justify your answer.

> **True.**
> p: Two is an even number. True   q: Three is an even number. False
> The disjunction p or q is true when any part is true.

2. Write a statement that is logically equivalent to the statement "If two sides of a triangle are congruent, the angles opposite those sides are congruent."
Identify the new statement as the converse, inverse, or contrapositive of the original statement.

> " If the angles opposite the two sides of a triangle are not congruent, those sides are not congruent."
> This is the contrapositive of the original statement.

IV. THEOREMS

1. The diagram below illustrates the construction of $\overleftrightarrow{PS}$ parallel to $\overleftrightarrow{RQ}$ through point P.

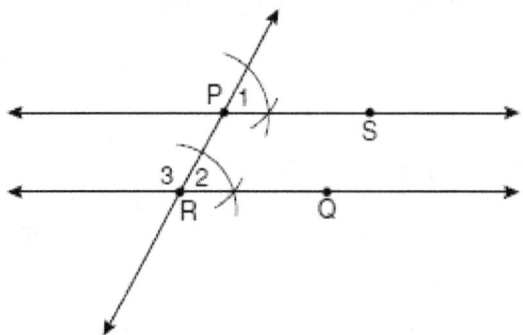

Which statement justifies this construction?

*(1) $m\angle 1 = m\angle 2$	(3) $\overline{PR} \cong \overline{RQ}$
(2) $m\angle 1 = m\angle 3$	(4) $\overline{PS} \cong \overline{RQ}$

Two lines are // if their corresponding angles are $\cong$.

2. Based on the diagram below, which statement is true?

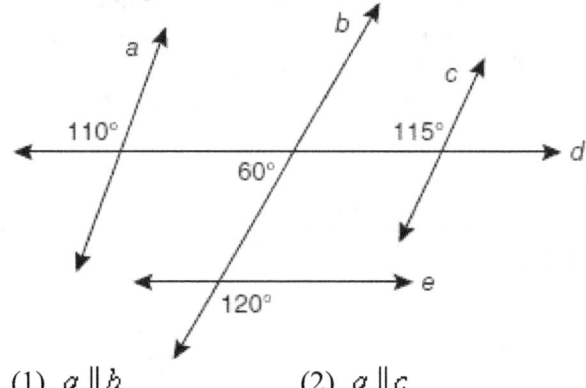

(1) $a \parallel b$ (2) $a \parallel c$ (3) $b \parallel c$ *(4) $d \parallel e$

Two lines are // if their alternate interior angles are $\cong$.

V. TRIANGLES

1. Juliann plans on drawing $\triangle ABC$, where the measure of $\angle A$ can range from 50° to 60° and the measure of $\angle B$ can range from 90° to 100°. Given these conditions, what is the correct range of measures possible for $\angle C$?
*(1) 20° to 40°   (2) 30° to 50°   (3) 80° to 90°   (4) 120° to 130°

> $m\angle A + m\angle B + m\angle C = 180$
> $180 - (60 + 100) = 20$, $180 - (50 + 90) = 40$

2. In an equilateral triangle, what is the difference between the sum of the exterior angles and the sum of the interior angles?
*(1) 180°   (2) 120°   (3) 90°   (4) 60°

> The sum of the exterior angles of any polygon is 360°. The sum of the interior angles of any triangle is 180°.
> $360 - 180 = 180$

3. In $\triangle ABC$, $m\angle A = x$, $m\angle B = 2x + 2$, and $m\angle C = 3x + 4$. What is the value of x?
*(1) 29   (2) 31   (3) 59   (4)61

> $x + 2x + 2 + 3x + 4 = 180$, $6x + 6 = 180$, $x = 29$

4. In the diagram below, $\triangle ABC$ is shown with $\overline{AC}$ extended through point D.

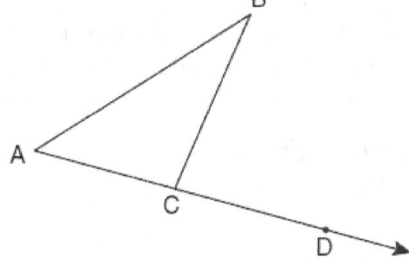

If $m\angle BCD = 6x + 2$, $m\angle BAC = 3x + 15$, and $m\angle ABC = 2x - 1$, what is the value of x?

*(1)12   (2) $14\frac{10}{11}$ (3) 16   (4) $18\frac{1}{9}$

> Exterior angle = the sum of 2 nonadjacent interior angles
> $6x + 2 = 3x + 15 + 2x - 1$, $6x + 2 = 5x + 14$, $x = 12$

5. In the diagram below of $\triangle ABC$, D is a point on $\overline{AB}$, $AC = 7$, $AD = 6$, and $BC = 18$.

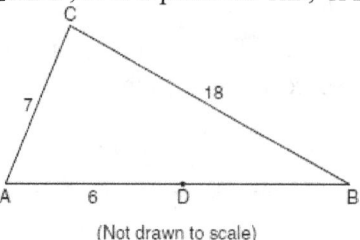

(Not drawn to scale)

The length of $\overline{DB}$ could be

(1) 5   *(2) 12   (3) 19   (4) 25

$$|s_1 - s_2| < s_3 < |s_1 + s_2|$$
$$|18 - 7| < AB < |18 + 7|, \qquad 11 < AB < 25$$
$$11 - 6 < DB < 25 - 6, \qquad 5 < DB < 19$$

6. In $\triangle ABC$, $m\angle A = 95$, $m\angle B = 50$, and $m\angle C = 35$. Which expression correctly relates the lengths of the sides of this triangle?

(1) $AB < BC < CA$ *(2) $AB < AC < BC$ (3) $AC < BC < AB$ (4) $BC < AC < AB$

$\angle C$ opposite AB , $\angle B$ opposite AC , $\angle A$ opposite BC

$m\angle C < m\angle B < m\angle A$, $AB < AC < BC$ In a triangle, larger angle is opposite longer side.

7. Side $\overline{PQ}$ of $\triangle PQR$ is extended through Q to point T. Which statement is *not* always true?

(1) $m\angle RQT > m\angle R$ (2) $m\angle RQT > m\angle P$ (3) $m\angle RQT = m\angle P + m\angle R$ *(4) $m\angle RQT > m\angle PQR$

The exterior angle of a triangle is greater than either nonadjacent interior angle, but not always greater than its adjacent interior angle. e.g. The obtuse angle of a triangle is greater than its adjacent interior angle.

8. Which set of numbers represents the lengths of the sides of a triangle?

(1) $\{5, 18, 13\}$ *(2) $\{6, 17, 22\}$ (3) $\{16, 24, 7\}$ (4) $\{26, 8, 15\}$

The sum of two shorter sides must be greater than the longest side.

9. Given $\triangle ABC$ with base $\overline{AFEDC}$, median $\overline{BF}$, altitude $\overline{BD}$, and $\overline{BE}$ bisects $\angle ABC$, which conclusion is valid?

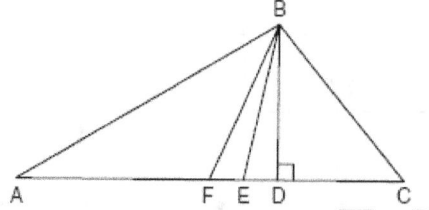

(1) $\angle FAB \cong \angle ABF$ (2) $\angle ABF \cong \angle CBD$ (3) $\overline{CE} \cong \overline{EA}$ *(4) $\overline{CF} \cong \overline{FA}$

$\overline{BF}$ is the median. F is the midpoint.

10. In which triangle do the three altitudes intersect outside the triangle?
(1) a right triangle   (2) an acute triangle   ***(3) an obtuse triangle**   (4) an equilateral triangle

11. In the diagram of $\triangle ABC$ below, Jose found centroid P by constructing the three medians. He measured $\overline{CF}$ and found it to be 6 inches.

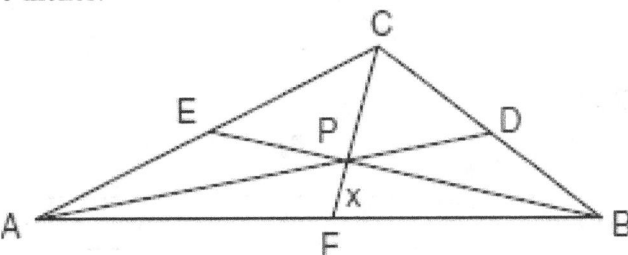

If $PF = x$, which equation can be used to find x?

(1) $x + x = 6$ ***(2) $2x + x = 6$**   (3) $3x + 2x = 6$ (4) $x + \dfrac{2}{3}x = 6$

The centroid divides each median in the ratio 2 to 1.

12. In the diagram of $\triangle ABC$ below, $\overline{AB} \cong \overline{AC}$. The measure of $\angle B$ is 40°.

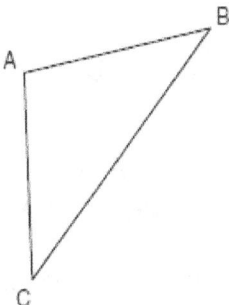

What is the measure of $\angle A$?
(1) 40°   (2) 50°   (3) 70°   ***(4) 100°**

$m\angle C = m\angle B = 40$, $m\angle A = 180 - 40 - 40 = 100$

13. In $\triangle ABC$, $\overline{AB} \cong \overline{BC}$. An altitude is drawn from B to $\overline{AC}$ and intersects $\overline{AC}$ at D. Which conclusion is *not* always true?
(1) $\angle ABD \cong \angle CBD$ (2) $\angle BDA \cong \angle BDC$ ***(3) $\overline{AD} \cong \overline{BD}$**   (4) $\overline{AD} \cong \overline{DC}$

To the base of an isosceles triangle, the altitude is also the angle bisector and median.

14. In the diagram below of $\triangle ADB$, m$\angle BDA = 90$, $AD = 5\sqrt{2}$, and $AB = 2\sqrt{15}$.

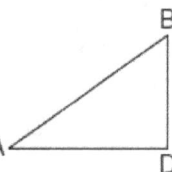

What is the length of $\overline{BD}$?

*(1) $\sqrt{10}$ (2) $\sqrt{20}$ (3) $\sqrt{50}$ (4) $\sqrt{110}$

$AB^2 = BD^2 + AD^2$, $BD = \sqrt{AB^2 - AD^2} = \sqrt{60 - 50} = \sqrt{10}$

15. In the diagram of $\triangle ABC$ and $\triangle DEF$ below, $\overline{AB} \cong \overline{DE}$, $\angle A \cong \angle D$, and $\angle B \cong \angle E$.

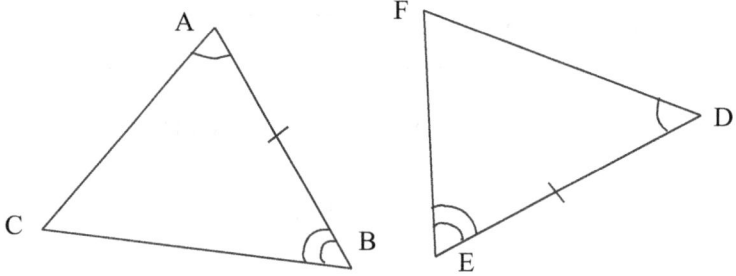

Which method can be used to prove $\triangle ABC \cong \triangle DEF$?

(1)   SSS   (2)   SAS   *(3)   ASA   (4)   HL

16. In the diagram of trapezoid $ABCD$ below, diagonals $\overline{AC}$ and $\overline{BD}$ intersect at E and $\triangle ABC \cong \triangle DCB$.

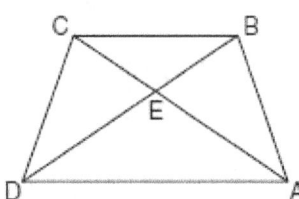

Which statement is true based on the given information?

(1) $\overline{AC} \cong \overline{BC}$ (2) $\overline{CD} \cong \overline{AD}$ (3) $\angle CDE \cong \angle BAD$ *(4) $\angle CDB \cong \angle BAC$

CPCTC: Corresponding Parts of Congruent Triangles are Congruent.

17. The diagonal $\overline{AC}$ is drawn in parallelogram $ABCD$. Which method can *not* be used to prove that $\triangle ABC \cong \triangle CDA$?

(1)   SSS   (2)   SAS   *(3)   SSA   (4)   ASA

SSA is not valid to prove two triangles $\cong$.

18. Two triangles are similar, and the ratio of each pair of corresponding sides is 2 : 1. Which statement regarding the two triangles is *not* true?
(1) Their areas have a ratio of 4 : 1.   (2) Their altitudes have a ratio of 2 : 1.
(3) Their perimeters have a ratio of 2 : 1.   ***(4) Their corresponding angles have a ratio of 2 : 1.**

Corresponding angles of two similar triangles are ≅ .

19. In the diagram below of $\triangle PRT$, Q is a point on $\overline{PR}$, S is a point on $\overline{TR}$, $\overline{QS}$ is drawn, and $\angle RPT \cong \angle RSQ$.

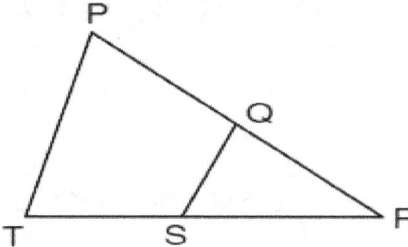

Which reason justifies the conclusion that $\triangle PRT \sim \triangle SRQ$?
*(1)   AA   (2)   ASA   (3)   SAS   (4)   SSS

20. In the diagram of $\triangle ABC$ and $\triangle EDC$ below, $\overline{AE}$ and $\overline{BD}$ intersect at C, and $\angle CAB \cong \angle CED$.

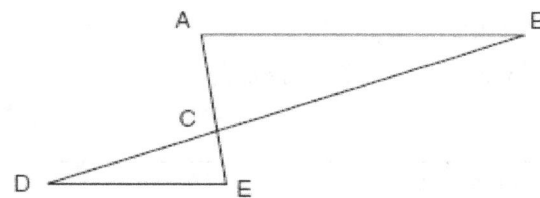

Which method can be used to show that $\triangle ABC$ must be similar to $\triangle EDC$?
(1)   SAS   ***(2)   AA**   (3)   SSS   (4)   HL

21. In $\triangle ABC$, point D is on $\overline{AB}$, and point E is on $\overline{BC}$ such that $\overline{DE} \parallel \overline{AC}$. If $DB = 2$, $DA = 7$, and $DE = 3$, what is the length of $\overline{AC}$?
(1)   8   (2)   9   (3)   10.5   ***(4)   13.5**

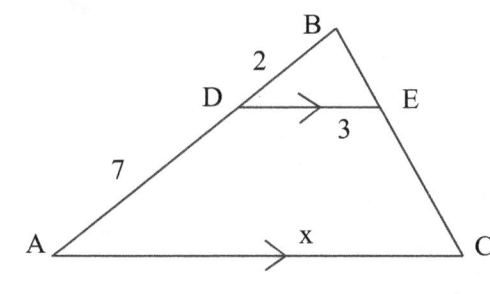

$$\frac{3}{x} = \frac{2}{2+7}$$

$$2x = 27$$

$$x = 13.5$$

22. In the diagram below of $\triangle ACT$, D is the midpoint of $\overline{AC}$, O is the midpoint of $\overline{AT}$, and G is the midpoint of $\overline{CT}$.

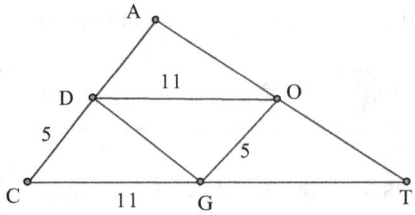

If $AC = 10$, $AT = 18$, and $CT = 22$, what is the perimeter of parallelogram $CDOG$?

(1)   21   (2)   25   *(3)   32   (4)   40

> $P = 5 + 5 + 11 + 11 = 32$

23. In the diagram below, $\overline{SQ}$ and $\overline{PR}$ intersect at T, $\overline{PQ}$ is drawn, and $\overline{PS} \parallel \overline{QR}$.

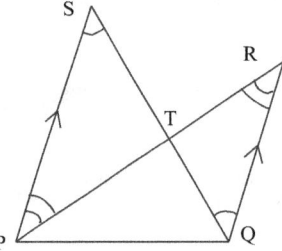

What technique can be used to prove that $\triangle PST \sim \triangle RQT$?

(1)   SAS   (2)   SSS   (3)   ASA   *(4)   AA

> If two lines are // , their alternate interior angles are $\cong$.

24. Given $\triangle ABC \sim \triangle DEF$ such that $\dfrac{AB}{DE} = \dfrac{3}{2}$. Which statement is *not* true?

(1) $\dfrac{BC}{EF} = \dfrac{3}{2}$ *(2) $\dfrac{m\angle A}{m\angle D} = \dfrac{3}{2}$ (3) $\dfrac{\text{area of } \triangle ABC}{\text{area of } \triangle DEF} = \dfrac{9}{4}$ (4) $\dfrac{\text{perimeter of } \triangle ABC}{\text{perimeter of } \triangle DEF} = \dfrac{3}{2}$

> Corresponding angles of two similar triangles are $\cong$.

25. In the diagram below, the length of the legs $\overline{AC}$ and $\overline{BC}$ of right triangle ABC are 6 cm and 8 cm, respectively. Altitude $\overline{CD}$ is drawn to the hypotenuse of $\triangle ABC$.

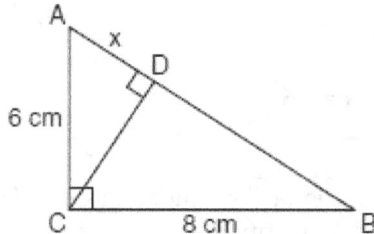

What is the length of $\overline{AD}$ to the *nearest tenth of a centimeter*?
*(1)   **3.6**   (2)   6.0   (3)   6.4   (4)   4.0

$$AB = \sqrt{6^2 + 8^2} = 10, \quad 6^2 = x \cdot 10, \quad 36 = 10x, \quad x = 3.6$$

26. In the diagram below of right triangle ACB, altitude $\overline{CD}$ is drawn to hypotenuse $\overline{AB}$.

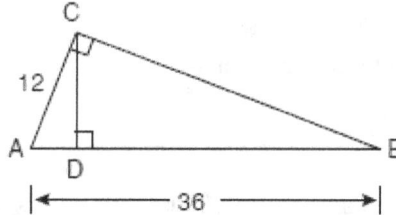

If $AB = 36$ and $AC = 12$, what is the length of $\overline{AD}$?
(1)   32   (2)   6   (3)   3   *(4)   **4**

$$12^2 = AD \cdot 36, \quad AD = \frac{144}{36} = 4$$

Show Work:

1. The degree measures of the angles of $\triangle ABC$ are represented by x, $3x$, and $5x - 54$. Find the value of x.

---

m∠A + m∠B + m∠C = 180
x + 3x + 5x -54 = 180 , 9x = 234 , **x = 26**

---

2. In the diagram below of $\triangle ABC$ with side $\overline{AC}$ extended through D, m∠A = 37 and m∠BCD = 117. Which side of $\triangle ABC$ is the longest side? Justify your answer.

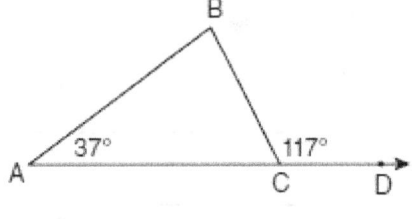

(Not drawn to scale)

---

m∠ACB = 180 - 117 = 63 , m∠B = 117 - 37 = 80 , m∠A = 37
In a triangle, the longest side is opposite the largest angle.
AC is the longest side.

---

3. In the diagram below of $\triangle TEM$, medians $\overline{TB}$, $\overline{EC}$, and $\overline{MA}$ intersect at D, and $TB = 9$.
Find the length of $\overline{TD}$.

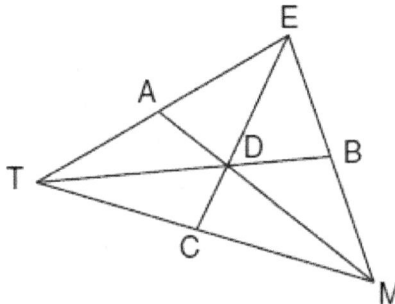

The centroid divides each median in the ratio 2 to 1.
$TD = \frac{2}{3} TB = \frac{2}{3} \cdot 9 = 6$

---

4. In $\triangle RST$, m$\angle RST = 46$ and $\overline{RS} \cong \overline{ST}$. Find m$\angle STR$.

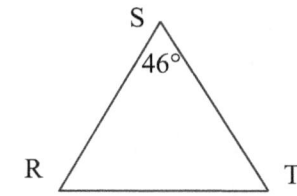

$$m\angle STR = \frac{180 - 46}{2} = \mathbf{67}$$

5. Given: $\triangle ABC$ and $\triangle EDC$, C is the midpoint of $\overline{BD}$ and $\overline{AE}$
Prove: $\overline{AB} \parallel \overline{DE}$

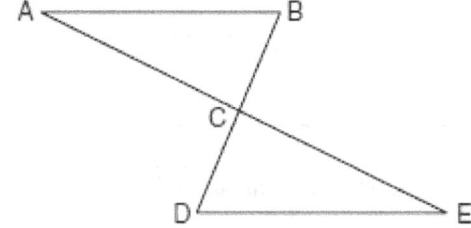

Statement	Reason
1. C is the midpoint of $\overline{BD}$ and $\overline{AE}$	1. Given
2. AC $\cong$ EC , BC $\cong$ DC	2. Midpoint divides a line segment into two $\cong$ segments.
3. $\angle$ACB $\cong$ $\angle$ECD	3. Vertical angles are $\cong$.
4. $\triangle$ABC $\cong$ $\triangle$EDC	4. SAS $\cong$
5. $\angle$B $\cong$ $\angle$D	5. CPCTC
6. AB $\parallel$ DE	6. Two lines are $\parallel$ if their alternate interior angles are $\cong$.

6. In the diagram of $\triangle ABC$ below, $AB = 10$, $BC = 14$, and $AC = 16$. Find the perimeter of the triangle formed by connecting the midpoints of the sides of $\triangle ABC$.

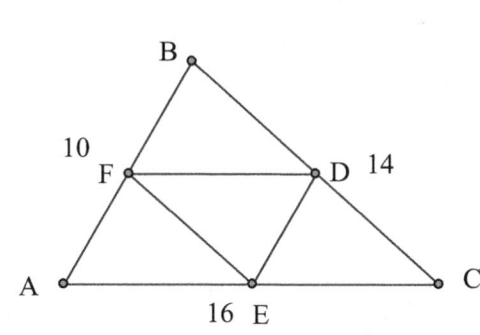

D, E, F are midpoints.

$$DE = \frac{1}{2}\,AB = 5$$

$$EF = \frac{1}{2}\,BC = 7$$

$$FD = \frac{1}{2}\,AC = 8$$

Perimeter = $5 + 7 + 8 = \mathbf{20}$

7. In the diagram below, $\triangle ABC \sim \triangle EFG$, $m\angle C = 4x + 30$, and $m\angle G = 5x + 10$.
Determine the value of x.

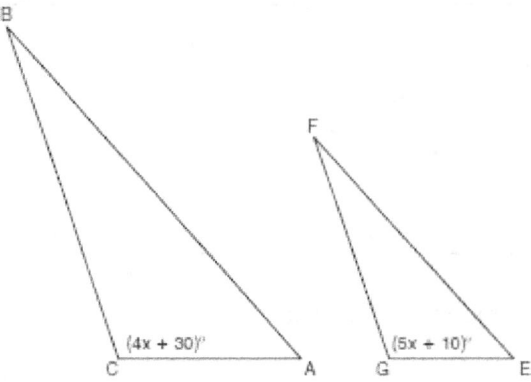

Corresponding angles of two similar triangles are $\cong$.
$5x + 10 = 4x + 30$, $x = \mathbf{20}$

8. In the diagram below of $\triangle ACD$, E is a point on $\overline{AD}$ and B is a point on $\overline{AC}$, such that $\overline{EB} \parallel \overline{DC}$.
If $AE = 3$, $ED = 6$, and $DC = 15$, find the length of $\overline{EB}$.

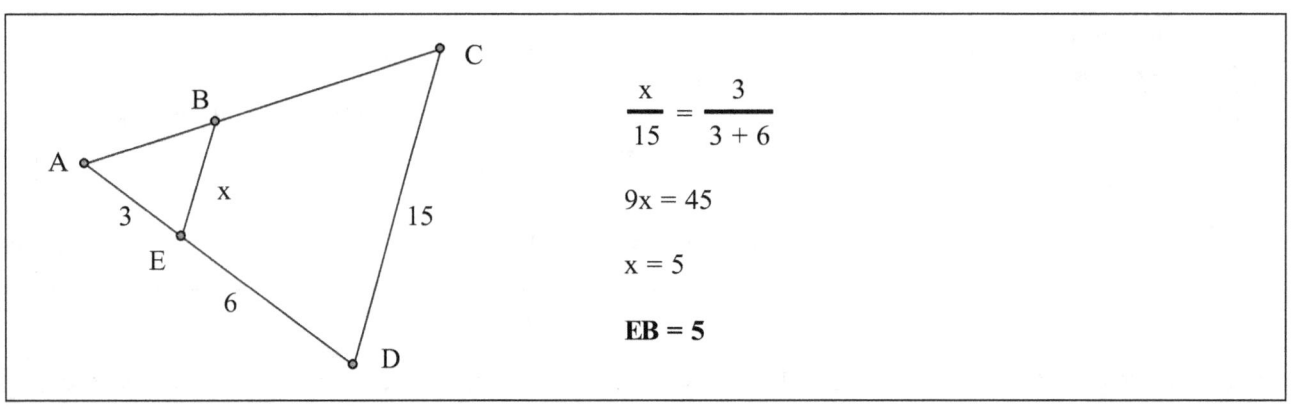

$$\frac{x}{15} = \frac{3}{3+6}$$

$$9x = 45$$

$$x = 5$$

EB = 5

9. In the diagram below of right triangle ACB, altitude $\overline{CD}$ intersects $\overline{AB}$ at D. If $AD = 3$ and $DB = 4$,
find the length of $\overline{CD}$ in simplest radical form.

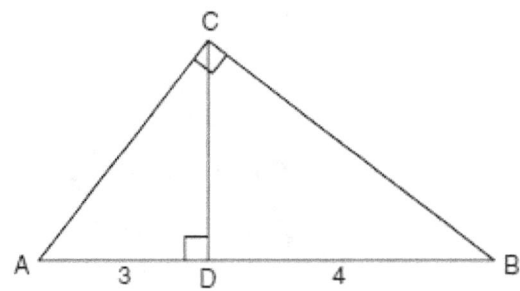

$CD^2 = 3 \cdot 4$, $CD = \sqrt{12} = \mathbf{2\sqrt{3}}$

VI. POLYGONS

1. In the diagram below of parallelogram $ABCD$ with diagonals $\overline{AC}$ and $\overline{BD}$, $m\angle 1 = 45$ and $m\angle DCB = 120$.

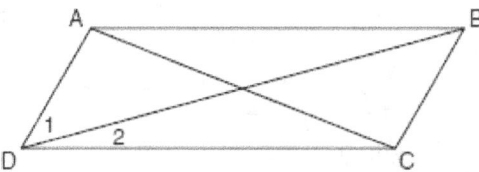

What is the measure of $\angle 2$?

*(1)   15°   (2)   30°   (3)   45°   (4)   60°

$\overline{AD} \parallel \overline{BC}$, $m\angle ADC = 180 - 120 = 60$, $m\angle 2 = 60 - 45 = 15$

---

2. A quadrilateral whose diagonals bisect each other and are perpendicular is a

*(1) rhombus   (3) trapezoid
(2) rectangle   (4) Parallelogram

3. In the diagram below of parallelogram $STUV$, $SV = x + 3$, $VU = 2x - 1$, and $TU = 4x - 3$.

What is the length of $\overline{SV}$?

*(1)   5   (2)   2   (3)   7   (4)   4

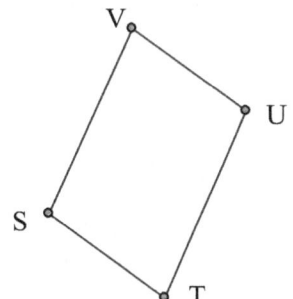

Opposite sides of a parallelogram are $\cong$.

TU = SV
4x - 3 = x + 3
3x = 6
x = 2
SV = x + 3 = 2 + 3 = **5**

---

4. Isosceles trapezoid $ABCD$ has diagonals $\overline{AC}$ and $\overline{BD}$. If $AC = 5x + 13$ and $BD = 11x - 5$, what is the value of x?

(1)   28   (2) $10\frac{3}{4}$ ***(3)   3**   (4) $\frac{1}{2}$

$AC = BD$, $5x + 13 = 11x - 5$, $6x = 18$, $x = 3$

---

5. In the diagram below of trapezoid *RSUT*, $\overline{RS} \parallel \overline{TU}$, *X* is the midpoint of $\overline{RT}$, and *V* is the midpoint of $\overline{SU}$.

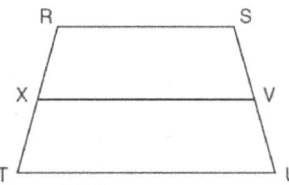

If *RS* = 30 and *XV* = 44, what is the length of $\overline{TU}$?
(1)   37   *(2)   **58**   (3)   74   (4)   118

$$\frac{30 + TU}{2} = 44 , \quad 30 + TU = 2 \bullet 44 , \quad TU = 58$$

6. What is the measure of an interior angle of a regular octagon?
(1)   45°   (2)   60°   (3)   120°   *(4)   **135°**

Octagon: n = 8 , $\frac{(n-2)180}{n} = \frac{(8-2)180}{8} = 135$

7. The pentagon in the diagram below is formed by five rays.

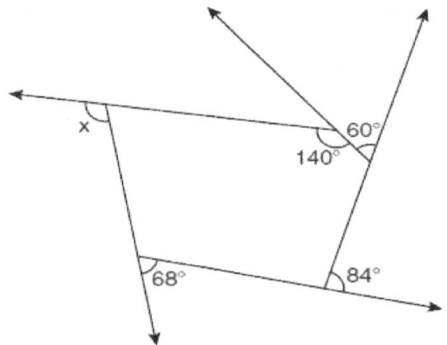

What is the degree measure of angle *x*?
(1)   72   (2)   96   *(3)   **108**   (4)   112

The supplement of 140° is 40°.
The sum of the exterior angles of any polygon is 360°.
x + 68 + 84 + 60 + 40 = 360 ,   x = 108

Show Work:

1. Given: Quadrilateral *ABCD*, diagonal $\overline{AFEC}$, $\overline{AE} \cong \overline{FC}$, $\overline{BF} \perp \overline{AC}$, $\overline{DE} \perp \overline{AC}$, $\angle 1 \cong \angle 2$
Prove: *ABCD* is a parallelogram.

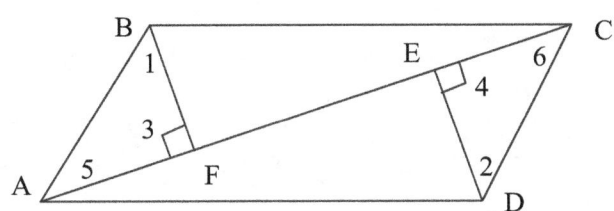

Statement	Reason
1. $\overline{AE} \cong \overline{FC}$, $\overline{BF} \perp \overline{AC}$, $\overline{DE} \perp \overline{AC}$, $\angle 1 \cong \angle 2$	1. Given
2. $\overline{FE} \cong \overline{FE}$	2. Reflexive property
3. AE - FE $\cong$ FC - FE or $\overline{AF} \cong \overline{CE}$	3. Subtraction postulate
4. $\angle 3$ and $\angle 4$ are right angles	4. $\perp$ lines form right angles
5. $\angle 3 \cong \angle 4$	5. Right angles are $\cong$
6. $\triangle ABF \cong \triangle CDE$	6. AAS $\cong$
7. $\overline{AB} \cong \overline{CD}$, $\angle 5 \cong \angle 6$	7. CPCTC
8. AB ‖ CD	8. Two lines are ‖ if their alternate interior angles $\cong$
9. ABCD is a parallelogram.	9. A quadrilateral is a parallelogram if it has one pair of opposite sides are $\cong$ and ‖

2. Given: *JKLM* is a parallelogram.
          $\overline{JM} \cong \overline{LN}$
          $\angle LMN \cong \angle LNM$
Prove: *JKLM* is a rhombus.

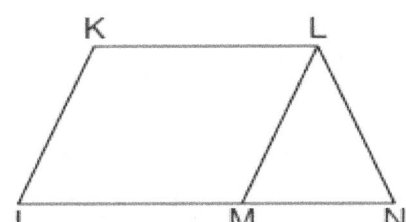

Statement	Reason
1. JKLM is a parallelogram, $\overline{JM} \cong \overline{LN}$, $\angle LMN \cong \angle LNM$	1. Given
2. $\overline{LN} \cong \overline{LM}$	2. If two angles of a $\triangle$ are $\cong$, their opposite sides are $\cong$
3. JM $\cong$ LM	3. Transitive property
4. JKLM is a rhombus	4. A rhombus is a parallelogram with two $\cong$ consecutive sides

3. In the diagram below of isosceles trapezoid $DEFG$, $\overline{DE} \parallel \overline{GF}$, $DE = 4x - 2$, $EF = 3x + 2$, $FG = 5x - 3$, and $GD = 2x + 5$. Find the value of x.

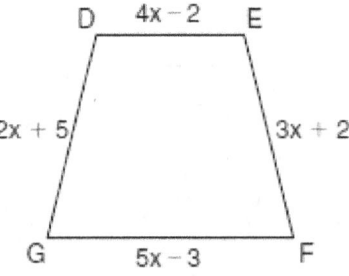

Isosceles trapezoid: nonparallel sides are $\cong$.
$EF = GD$, $\quad 3x + 2 = 2x + 5$, $\quad$ **x = 3**

4. In the diagram below, quadrilateral $STAR$ is a rhombus with diagonals $\overline{SA}$ and $\overline{TR}$ intersecting at E. $ST = 3x + 30$, $SR = 8x - 5$, $SE = 3z$, $TE = 5z + 5$, $AE = 4z - 8$, $m\angle RTA = 5y - 2$, and $m\angle TAS = 9y + 8$. Find SR, RT, and $m\angle TAS$.

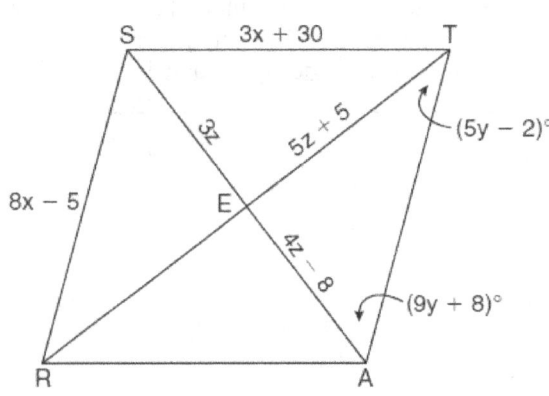

(1).   Rhombus: 4 sides are $\cong$.   SR = ST
       8x - 5 = 3x + 30
          5x = 35 ,   x = 7
       SR = 8x - 5 = 8•7 - 5 = **51**
(2).   Rhombus: Diagonals bisect each other. AE = SE
       4z - 8 = 3z
          z = 8
       RT = 2TE = 2(5z + 5) = 2(5•8 + 5) = **90**
(3).   Rhombus: Diagonals are $\perp$.
       $\angle RTA$ and $\angle TAS$ are supplementary.
       5y - 2 + 9y + 8 = 90
           14y = 84
             y = 6
       m$\angle$TAS = 9y + 8 = 9•6 + 8 = **62**

VII. COORDINATE GEOMETRY

1. Line segment AB has endpoints $A(2,-3)$ and $B(-4, 6)$. What are the coordinates of the midpoint of $\overline{AB}$?

(1) $(-2, 3)$ *(2) $\left(-1, 1\frac{1}{2}\right)$ (3) $(-1, 3)$ (4) $\left(3, 4\frac{1}{2}\right)$

$$M(\bar{x}, \bar{y}) = M(\frac{x_1 + x_2}{2}, \frac{y_1 + y_2}{2}) = M(\frac{2 - 4}{2}, \frac{-3 + 6}{2}) = M(-1, 1\frac{1}{2})$$

2. Square $LMNO$ is shown in the diagram below.

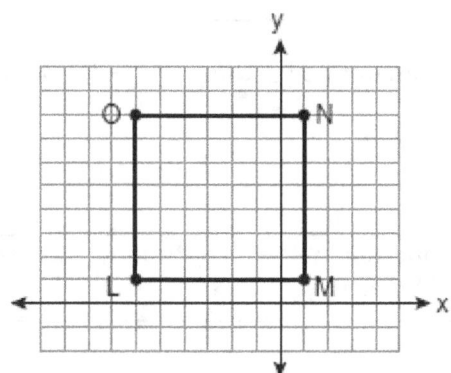

What are the coordinates of the midpoint of diagonal $\overline{LN}$?

(1) $\left(4\frac{1}{2}, -2\frac{1}{2}\right)$ (2) $\left(-3\frac{1}{2}, 3\frac{1}{2}\right)$ (3) $\left(-2\frac{1}{2}, 3\frac{1}{2}\right)$ *(4) $\left(-2\frac{1}{2}, 4\frac{1}{2}\right)$

L(-6, 1), N(1, 8)

$$M(\bar{x}, \bar{y}) = M(\frac{x_1 + x_2}{2}, \frac{y_1 + y_2}{2}) = M(\frac{-6 + 1}{2}, \frac{1 + 8}{2}) = M(-2\frac{1}{2}, 4\frac{1}{2})$$

3. The endpoints of $\overline{CD}$ are $C(-2,-4)$ and $D(6, 2)$. What are the coordinates of the midpoint of $\overline{CD}$?
(1) $(2, 3)$ *(2) $(2, -1)$ (3) $(4, -2)$ (4) $(4, 3)$

$$M(\bar{x}, \bar{y}) = M(\frac{x_1 + x_2}{2}, \frac{y_1 + y_2}{2}) = M(\frac{-2 + 6}{2}, \frac{-4 + 2}{2}) = M(2, -1)$$

4. If the endpoints of $\overline{AB}$ are $A(-4, 5)$ and $B(2, -5)$, what is the length of $\overline{AB}$?

*(1) $2\sqrt{34}$ (2)   2   (3) $\sqrt{61}$ (4)   8

$$AB = \sqrt{(x_2 - x_1)^2 + (y_2 - y_1)^2} = \sqrt{(-4-2)^2 + (5+5)^2} = \sqrt{36 + 100} = \sqrt{136} = \mathbf{2\sqrt{34}}$$

5. What is the distance between the points $(-3, 2)$ and $(1, 0)$?

(1) $2\sqrt{2}$ (2) $2\sqrt{3}$ (3) $5\sqrt{2}$ ***(4)** $2\sqrt{5}$

$$d = \sqrt{(x_2 - x_1)^2 + (y_2 - y_1)^2} = \sqrt{(-3-1)^2 + (2-0)^2} = \sqrt{16 + 4} = \sqrt{20} = \mathbf{2\sqrt{5}}$$

6. What is the equation of a line that passes through the point $(-3, -11)$ and is parallel to the line whose equation is $2x - y = 4$?

(1) $y = 2x + 5$ ***(2)** $y = 2x - 5$ (3) $y = \frac{1}{2}x + \frac{25}{2}$ (4) $y = -\frac{1}{2}x - \frac{25}{2}$

Rewrite the given line $2x - y = 4$ in slope and y-intercept form: $y = 2x - 4$, slope is 2.
Test the point (-3, -11) in Eq. (1) and Eq. (2).

7. The lines $3y + 1 = 6x + 4$ and $2y + 1 = x - 9$ are

(1)   parallel (2)   perpendicular (3)   the same line   ***(4)neither parallel nor perpendicular**

Rewrite the equations in slope and y-intercept form.
$3y + 1 = 6x + 4$ is $y = 2x + 1$, $m = 2$; $2y + 1 = x - 9$ is $y = \frac{1}{2}x - 5$, $m = \frac{1}{2}$

8. What is the slope of a line perpendicular to the line whose equation is $5x + 3y = 8$?

(1) $\frac{5}{3}$ ***(2)** $\frac{3}{5}$ (3) $-\frac{3}{5}$ (4) $-\frac{5}{3}$

Rewrite the equation in slope and y-intercept form.
$3y = -5x + 8$, $y = -\frac{5}{3}x + \frac{8}{3}$, $m = -\frac{5}{3}$.
The slope of the perpendicular line is the negative reciprocal $\frac{3}{5}$.

9. What is an equation of the line that passes through the point $(-2, 5)$ and is perpendicular to the line whose equation is $y = \frac{1}{2}x + 5$?

(1) $y = 2x + 1$ ***(2)** $y = -2x + 1$ (3) $y = 2x + 9$ (4) $y = -2x - 9$

The slope of the perpendicular line is the negative reciprocal -2.
Test the point (-2, 5) in Eq. (2) and Eq. (4).

10. Which equation represents a line perpendicular to the line whose equation is $2x + 3y = 12$?

(1) $6y = -4x + 12$ *(2) $2y = 3x + 6$ (3) $2y = -3x + 6$ (4) $3y = -2x + 12$

Rewrite the equations in slope and y-intercept form.

The given line: $2x + 3y = 12$, $3y = -2x + 12$, $y = -\frac{2}{3}x + 4$, $m = -\frac{2}{3}$

Eq. (2): $2y = 3x + 6$, $y = \frac{3}{2}x + 3$, $m = \frac{3}{2}$, which is the negative reciprocal of $m = -\frac{2}{3}$.

---

11. What is the equation of a line that is parallel to the line whose equation is $y = x + 2$?

(1) $x + y = 5$ (2) $2x + y = -2$ *(3) $y - x = -1$ (4) $y - 2x = 3$

Rewrite the equation in slope and y-intercept form.

Eq. (3) is $y = x - 1$ which has the same slope as the given Eq.

---

12. What is the slope of a line perpendicular to the line whose equation is $y = -\frac{2}{3}x - 5$?

(1) $-\frac{3}{2}$ (2) $-\frac{2}{3}$ (3) $\frac{2}{3}$ *(4) $\frac{3}{2}$

The slopes of two perpendicular lines are negative reciprocals.

---

13. Which equation represents a line parallel to the line whose equation is $2y - 5x = 10$?

(1) $5y - 2x = 25$ (2) $5y + 2x = 10$ *(3) $4y - 10x = 12$ (4) $2y + 10x = 8$

Rewrite the equations in slope and y-intercept form.

The given line: $2y - 5x = 10$, $2y = 5x + 10$, $y = \frac{5}{2}x + 5$, $m = \frac{5}{2}$

Eq. (3): $4y - 10x = 12$, $4y = 10x + 12$, $y = \frac{5}{2}x + 3$, $m = \frac{5}{2}$

The slopes of two parallel lines are equal.

---

14. What is an equation of the line that contains the point $(3, -1)$ and is perpendicular to the line whose equation is $y = -3x + 2$?

(1) $y = -3x + 8$ (2) $y = -3x$ (3) $y = \frac{1}{3}x$ *(4) $y = \frac{1}{3}x - 2$

The slope of the perpendicular line is the negative reciprocal $\frac{1}{3}$.

Test the point $(3, -1)$ in Eq. (3) and Eq. (4).

---

15. What is the slope of a line that is perpendicular to the line whose equation is $3x + 4y = 12$?

(1) $\frac{3}{4}$ (2) $-\frac{3}{4}$ *(3) $\frac{4}{3}$ (4) $-\frac{4}{3}$

Rewrite the given line in slope and y-intercept form.

$3x + 4y = 12$, $4y = -3x + 12$, $y = -\frac{3}{4}x + 3$, $m = -\frac{3}{4}$, the negative reciprocal is $\frac{4}{3}$

---

16. Two lines are represented by the equations $-\frac{1}{2}y = 6x + 10$ and $y = mx$. For which value of m will the lines be parallel?

*(1)   **-12**   (2)   -3   (3)   3   (4)   12

Parallel lines have the same slope.

Rewrite the Eq. $-\frac{1}{2}y = 6x + 10$ in slope and y-intercept form.

$y = -12x - 20$, $m = -12$

---

17. The vertices of $\triangle ABC$ are $A(-1, -2)$, $B(-1, 2)$ and $C(6, 0)$. Which conclusion can be made about the angles of $\triangle ABC$?

*(1) $m\angle A = m\angle B$ (2) $m\angle A = m\angle C$ (3) $m\angle ACB = 90$ (4) $m\angle ABC = 60$

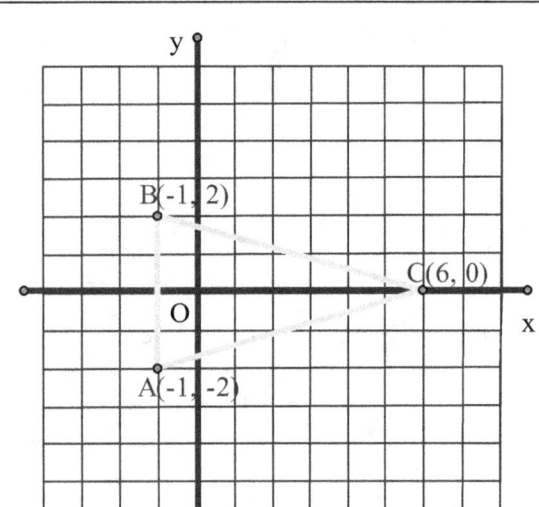

$\triangle ABC$ is symmetric about x-axis.

Show Work:

1. The endpoints of $\overline{PQ}$ are $P(-3, 1)$ and $Q(4, 25)$. Find the length of $\overline{PQ}$.

$$PQ = \sqrt{(x_2 - x_1)^2 + (y_2 - y_1)^2} = \sqrt{(4 + 3)^2 + (25 - 1)^2} = \sqrt{7^2 + 24^2} = \sqrt{625} = \mathbf{25}$$

2. Triangle ABC has coordinates $A(-6, 2)$, $B(-3, 6)$, and $C(5, 0)$. Find the perimeter of the triangle. Express your answer in simplest radical form.

Use distance formula: $d = \sqrt{(x_2 - x_1)^2 + (y_2 - y_1)^2}$

$AB = \sqrt{(-3 + 6)^2 + (6 - 2)^2} = \sqrt{9 + 16} = \sqrt{25} = 5$

$BC = \sqrt{(5 + 3)^2 + (0 - 6)^2} = \sqrt{64 + 36} = \sqrt{100} = 10$

$CA = \sqrt{(5 + 6)^2 + (0 - 2)^2} = \sqrt{121 + 4} = \sqrt{125} = 5\sqrt{5}$

Perimeter $= AB + BC + CA = 5 + 10 + 5\sqrt{5} = \mathbf{15 + 5\sqrt{5}}$

3. In the diagram below of circle C, $\overline{QR}$ is a diameter, and $Q(1, 8)$ and $C(3.5, 2)$ are points on a coordinate plane. Find and state the coordinates of point R.

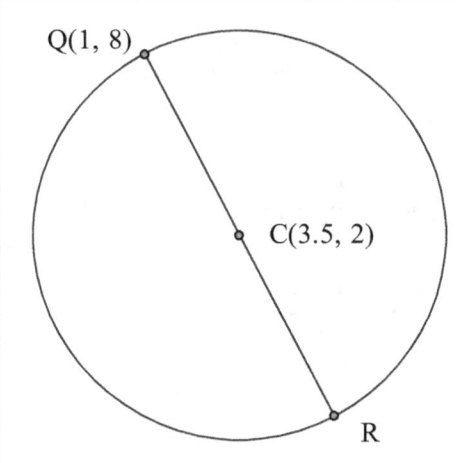

Q(1, 8)

C(3.5, 2)

R

Use midpoint formular: $\bar{x} = \dfrac{x_1 + x_2}{2}$, $\bar{y} = \dfrac{y_1 + y_2}{2}$

C(3.5, 2) is the midpoint.

$3.5 = \dfrac{1 + x_2}{2}$

$7 = 1 + x_2$

$x_2 = 6$

$2 = \dfrac{8 + y_2}{2}$

$4 = 8 + y_2$

$y_2 = -4$

R(6, -4)

4. Find an equation of the line passing through the point $(5, 4)$ and parallel to the line whose equation is $2x + y = 3$.

Find the slope of the given line $2x + y = 3$, $y = -2x + 3$, $m = -2$.
Method 1: Using the Point-Slope form $y - y_1 = m(x - x_1)$
$\qquad$ $y - 4 = -2(x - 5)$
Method 2: Using the Slope-Intercept form $y = mx + b$
$\qquad$ $y = -2x + b$, plug in $(5, 4)$
$\qquad$ $4 = -2(5) + b$, $b = 14$
$\qquad$ $y = -2x + 14$

5. Write an equation of the line that passes through the point $(6, -5)$ and is parallel to the line whose equation is $2x - 3y = 11$.

Find the slope of the given line $2x - 3y = 11$, $-3y = -2x + 11$, $y = \frac{2}{3}x - \frac{11}{3}$, $m = \frac{2}{3}$.
 Using the Point-Slope form $y - y_1 = m(x - x_1)$
$\qquad$ $y + 5 = \frac{2}{3}(x - 6)$

6. On the set of axes below, graph and label $\triangle DEF$ with vertices at $D(-4, -4)$, $E(-2, 2)$, and $F(8, -2)$. If G is the midpoint of $\overline{EF}$ and H is the midpoint of $\overline{DF}$, state the coordinates of G and H and label each point on your graph. Explain why $\overline{GH} \parallel \overline{DE}$.

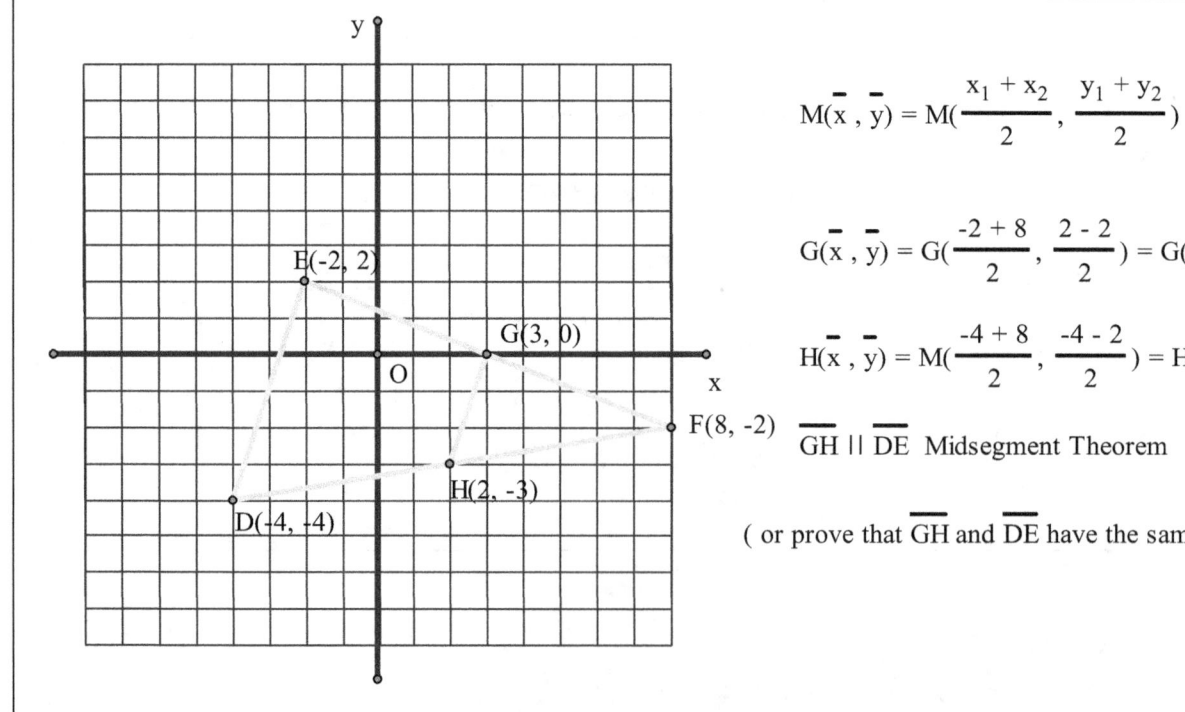

$$M(\overline{x}, \overline{y}) = M(\frac{x_1 + x_2}{2}, \frac{y_1 + y_2}{2})$$

$$G(\overline{x}, \overline{y}) = G(\frac{-2 + 8}{2}, \frac{2 - 2}{2}) = G(3, 0)$$

$$H(\overline{x}, \overline{y}) = M(\frac{-4 + 8}{2}, \frac{-4 - 2}{2}) = H(2, -3)$$

$\overline{GH} \parallel \overline{DE}$ Midsegment Theorem

(or prove that $\overline{GH}$ and $\overline{DE}$ have the same slope)

VIII. CIRCLE

1. In the diagram below of circle O, chords $\overline{AD}$ and $\overline{BC}$ intersect at E, $m\overarc{AC} = 87$, and $m\overarc{BD} = 35$. What is the degree measure of $\angle CEA$?

(1)　　87　　　　　　　　*(2)　61　　　　　　　(3)　　43.5　　　　　　(4)　　26

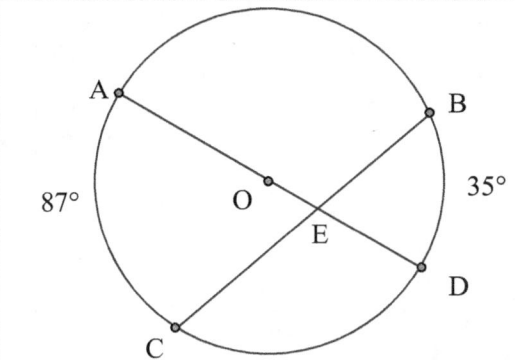

Measure of the Chord-Chord angle:

$$m\angle CEA = \frac{1}{2}(m\overarc{AC} + m\overarc{BD})$$
$$= \frac{1}{2}(87 + 35)$$
$$= 61$$

2. In the diagram below, $\overline{PS}$ is a tangent to circle O at point S, $\overline{PQR}$ is a secant, $PS = x$, $PQ = 3$, and $PR = x + 18$.

What is the length of $\overline{PS}$?

(1)　　6　　　　　　　　*(2)　9　　　　　　　(3)　　3　　　　　　　(4)　　27

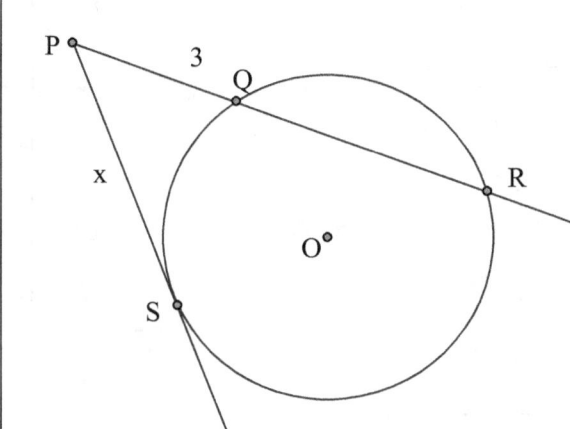

$PS^2 = PQ \bullet PR$

$x^2 = 3(x + 18)$

$x^2 = 3x + 54$

$x^2 - 3x - 54 = 0$

$(x - 9)(x + 6) = 0$

$\mathbf{x = 9}$　　　($x = -6$ rejected)

3. In the diagram below, tangent $\overline{AB}$ and secant $\overline{ACD}$ are drawn to circle O from an external point A, $AB = 8$, and $AC = 4$.

What is the length of $\overline{CD}$?

(1)   16   (2)   13   *(3)   **12**   (4)   10

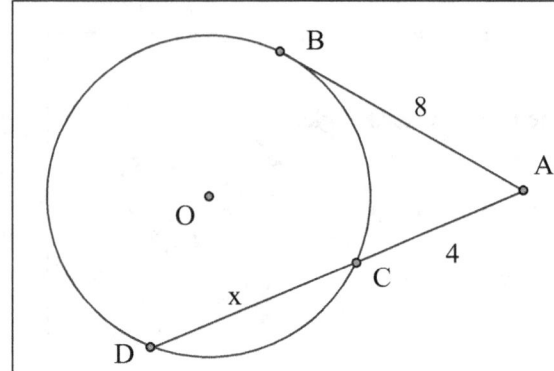

$$AB^2 = AC(AC + CD)$$

$$8^2 = 4(4 + x)$$

$$64 = 16 + 4x$$

$$48 = 4x$$

$$x = 12$$

4. In the diagram of circle O below, chord $\overline{AB}$ intersects chord $\overline{CD}$ at E, $DE = 2x + 8$, $EC = 3$, $AE = 4x - 3$, and $EB = 4$.

What is the value of x?

(1)   1   *(2)   **3.6**   (3)   5   (4)   10.25

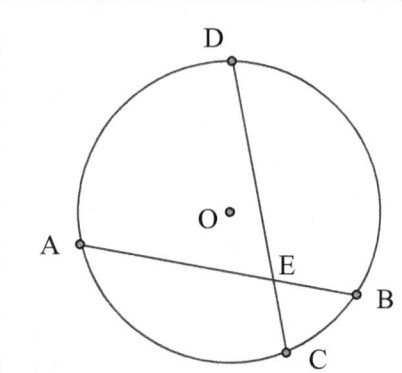

$$AE \bullet EB = DE \bullet EC$$

$$(4x - 3) \bullet 4 = (2x + 8) \bullet 3$$

$$16x - 12 = 6x + 24$$

$$10x = 36$$

$$x = 3.6$$

5. In the diagram below, tangent $\overline{PA}$ and secant $\overline{PBC}$ are drawn to circle O from external point P.

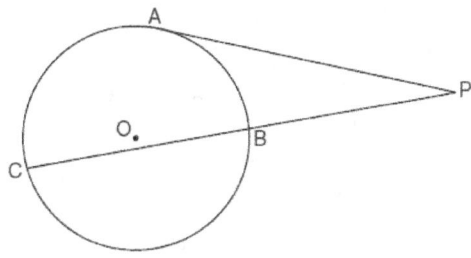

If $PB = 4$ and $BC = 5$, what is the length of $\overline{PA}$?

(1)   20   (2)   9   (3)   8   *(4)   6

$PA^2 = PB(PB + BC)$, $PA^2 = 4(4 + 5) = 36$, $PA = 6$

6. In the diagram below, circle O has a radius of 5, and $CE = 2$. Diameter $\overline{AC}$ is perpendicular to chord $\overline{BD}$ at E.

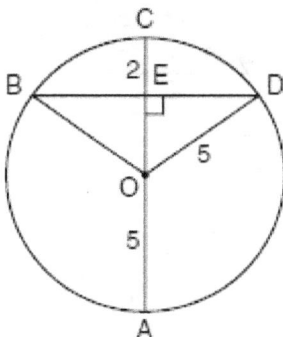

What is the length of $\overline{BD}$?

(1)   12   (2)   10   *(3)   8   (4)   4

OE = OC − CE = 5 − 2 = 3
Pythagorean Triple: ED = 4 , BD = 2ED = 8   ($ED = \sqrt{5^2 - 3^2} = 4$)

7. In the diagram of circle O below, chords $\overline{AB}$ and $\overline{CD}$ are parallel, and $\overline{BD}$ is a diameter of the circle.

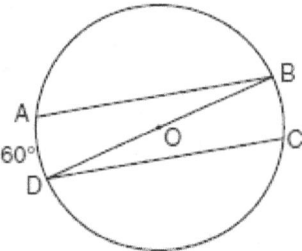

If $m\overset{\frown}{AD} = 60$, what is $m\angle CDB$?

(1)   20   *(2)   30   (3)   60   (4)   120

$$\overline{AB} \parallel \overline{DC}, \quad m\overset{\frown}{AD} = m\overset{\frown}{BC} = 60, \quad m\angle CDB = \frac{1}{2}m\overset{\frown}{BC} = \frac{1}{2} \cdot 60 = 30$$

8. In the diagram of circle O below, chord $\overline{CD}$ is parallel to diameter $\overline{AOB}$ and $m\overset{\frown}{AC} = 30$.

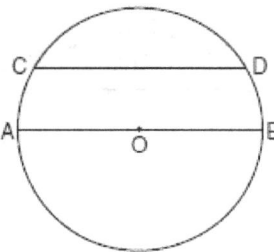

What is $m\overset{\frown}{CD}$?

(1)   150   * (2)   120   (3)   100   (4)   60

$$\overline{CD} \parallel \overline{AOB}, \quad m\overset{\frown}{AC} = m\overset{\frown}{BD} = 30, \quad m\overset{\frown}{CD} = 180 - 30 - 30 = 120 \quad \text{(semicircle 180°)}$$

9. In the diagram below, $\triangle ABC$ is inscribed in circle P. The distances from the center of circle P to each side of the triangle are shown.

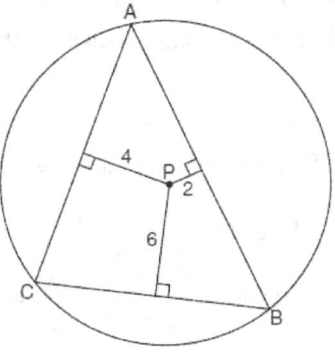

Which statement about the sides of the triangle is true?

*(1) $AB > AC > BC$ (2) $AB < AC$ and $AC > BC$ (3) $AC > AB > BC$ (4) $AC = AB$ and $AB > BC$

The length of the chord increases when the distance from the center decreases.

10. How many common tangent lines can be drawn to the two externally tangent circles shown below?

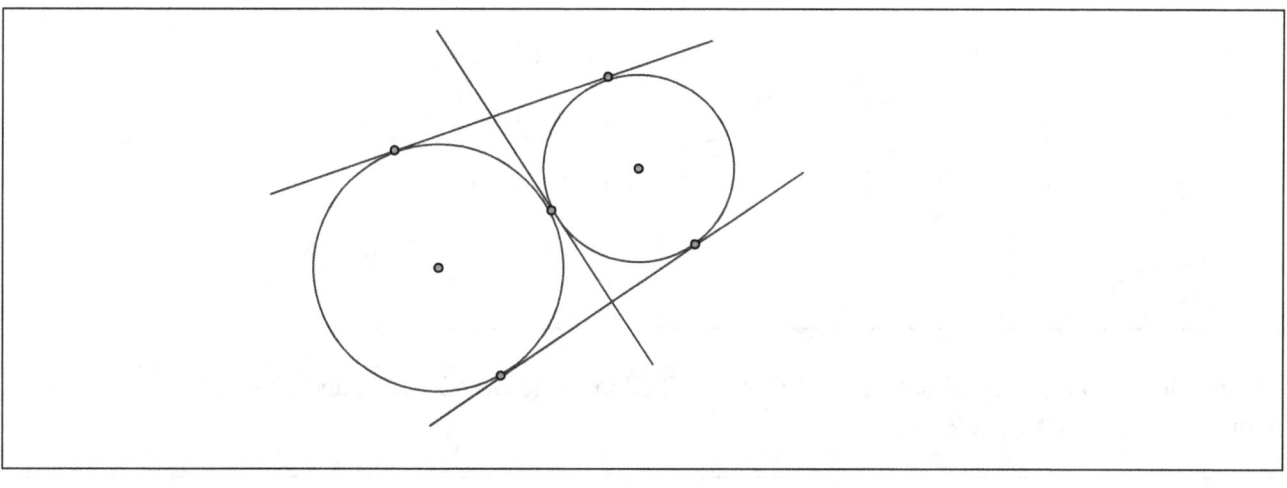

(1)   1   (2)   2   *(3)   3   (4)   4

Show Work:

1. In the diagram below of circle O, chords $\overline{DF}$, $\overline{DE}$, $\overline{FG}$, and $\overline{EG}$ are drawn such that $m\overarc{DF}:m\overarc{FE}:m\overarc{EG}:m\overarc{GD} = 5:2:1:7$. Identify one pair of inscribed angles that are congruent to each other and give their measure.

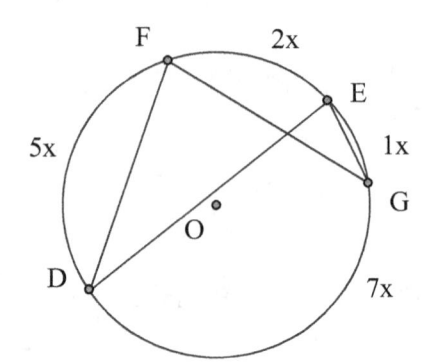

Two inscribed angles intercepting the same arc are $\cong$.

$$\angle D \cong \angle G$$

Find each measure of the arc around the circle first.
$x + 2x + 5x + 7x = 360$
$15x = 360$, $x = 24$
$m\overarc{FE} = 2x = 2 \cdot 24 = 48$
$m\angle D = m\angle G = \dfrac{1}{2} m\overarc{FE} = \dfrac{1}{2} \cdot 48 = \mathbf{24}$

2. In the diagram below, circles X and Y have two tangents drawn to them from external point T. The points of tangency are C, A, S, and E. The ratio of TA to AC is $1:3$. If $TS = 24$, find the length of $\overline{SE}$.

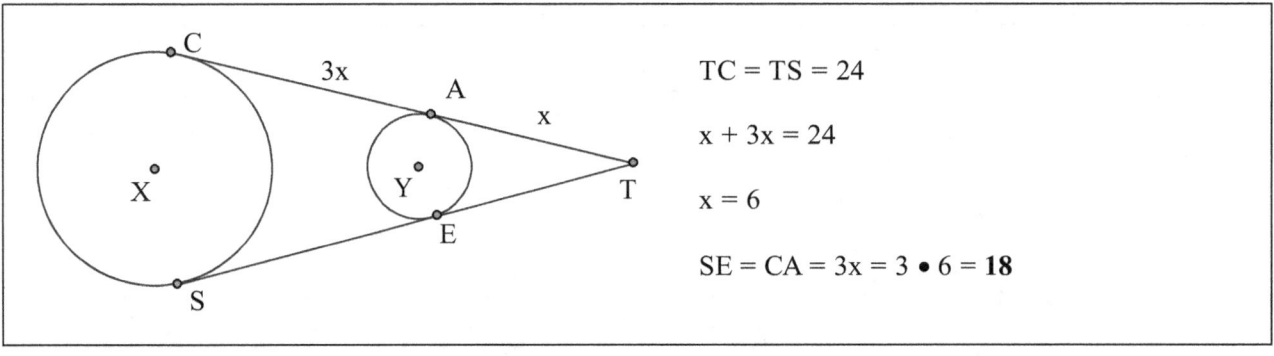

$TC = TS = 24$

$x + 3x = 24$

$x = 6$

$SE = CA = 3x = 3 \cdot 6 = \mathbf{18}$

3. In the diagram below, quadrilateral $ABCD$ is inscribed in circle O, $\overline{AB} \parallel \overline{DC}$, and diagonals $\overline{AC}$ and $\overline{BD}$ are drawn. Prove that $\triangle ACD \cong \triangle BDC$.

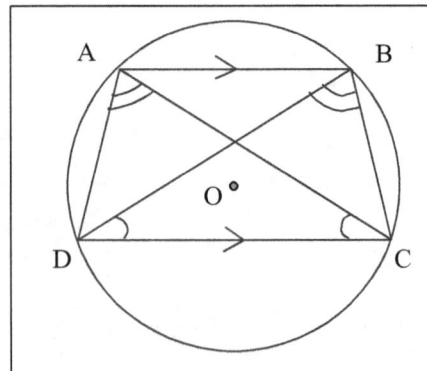

Statement	**Reason**
1. $\overline{AB} \parallel \overline{DC}$	1. Given
2. $\overarc{AD} \cong \overarc{BC}$	2. In a circle, $\parallel$ chords intercept $\cong$ arcs
3. $\angle ACD \cong \angle BDC$	3. Inscribed angles are $\cong$ if they intercept $\cong$ arcs
4. $\angle DAC \cong \angle CBD$	4. Inscribed angles are $\cong$ if they intercept the same arc
5. $\overline{DC} \cong \overline{DC}$	5. Reflexive property
6. $\triangle ACD \cong \triangle BDC$	6. AAS $\cong$

IX. CONSTRUCTIONS AND LOCI

1. The diagram below shows the construction of the perpendicular bisector of $\overline{AB}$.

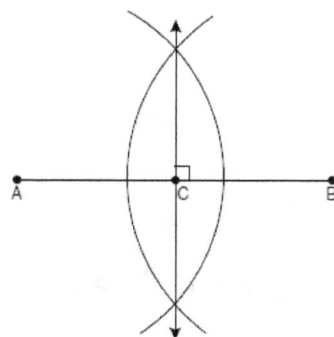

Which statement is *not* true?

(1) $AC = CB$ (2) $CB = \frac{1}{2} AB$ *(3) $AC = 2AB$ (4) $AC + CB = AB$

2. Which illustration shows the correct construction of an angle bisector?

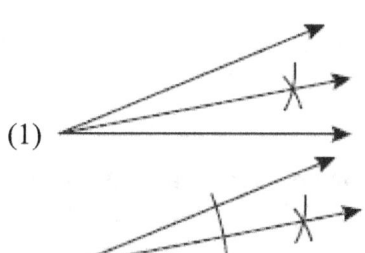

(1)   (2)

*(3)   (4)

3. The diagram below shows the construction of the bisector of $\angle ABC$.

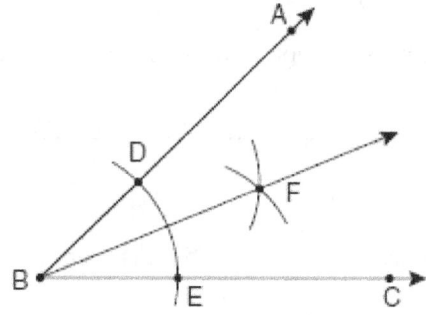

Which statement is *not* true?

(1) $m\angle EBF = \frac{1}{2} m\angle ABC$ (2) $m\angle DBF = \frac{1}{2} m\angle ABC$ *(3) $m\angle EBF = m\angle ABC$ (4) $m\angle DBF = m\angle EBF$

4. Based on the construction below, which statement must be true?

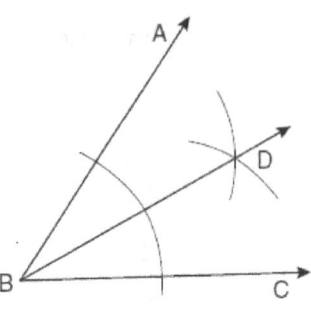

(1) $m\angle ABD = \frac{1}{2} m\angle CBD$ *(2) $m\angle ABD = m\angle CBD$ (3) $m\angle ABD = m\angle ABC$ (4) $m\angle CBD = \frac{1}{2} m\angle ABD$

5. Which geometric principle is used to justify the construction below?

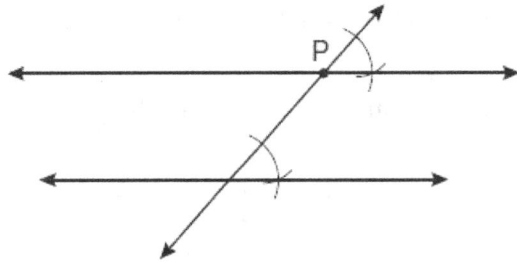

(1) A line perpendicular to one of two parallel lines is perpendicular to the other.

(2) Two lines are perpendicular if they intersect to form congruent adjacent angles.

(3) When two lines are intersected by a transversal and alternate interior angles are congruent, the lines are parallel.

*(4) **When two lines are intersected by a transversal and the corresponding angles are congruent, the lines are parallel.**

6. In a coordinate plane, how many points are both 5 units from the origin and 2 units from the x-axis?
(1)   1   (2)   2   (3)   3   *(4)   4

Locus 1

Locus 2

Locus 2

Locus 1: 5 units from the origin is a circle.

Locus 2: 2 units from the x-axis is a pair of parallel lines.

4 points of intersection.

7. Towns A and B are 16 miles apart. How many points are 10 miles from town A and 12 miles from town B?

(1)   1   *(2)   2   (3)   3   (4)   0

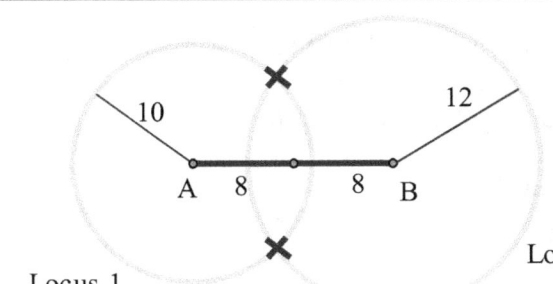

2 points of intersection

Locus 2

8. What are the center and radius of a circle whose equation is $(x - A)^2 + (y - B)^2 = C$?

(1)   center = (A, B); radius = C *(3)   center = (A, B); **radius** = $\sqrt{C}$

(2)   center = $(-A, -B)$; radius = C (4)   center = $(-A, -B)$; radius = $\sqrt{C}$

The center-radius form of a circle is $(x - h)^2 + (y - k)^2 = r^2$
Compare: $A = h$, $B = k$ and $r^2 = C$, $r = \sqrt{C}$

9. The diameter of a circle has endpoints at $(-2, 3)$ and $(6, 3)$. What is an equation of the circle?

*(1) $(x - 2)^2 + (y - 3)^2 = 16$ (3) $(x + 2)^2 + (y + 3)^2 = 16$

(2) $(x - 2)^2 + (y - 3)^2 = 4$ (4) $(x + 2)^2 + (y + 3)^2 = 4$

The center of the circle is the midpoint of the diameter.
$$(\bar{x}, \bar{y}) = (\frac{x_1 + x_2}{2}, \frac{y_1 + y_2}{2}) = (\frac{-2 + 6}{2}, \frac{3 + 3}{2}) = (2, 3)$$
The radius is the distance between the center $(2, 3)$ and any point of the circle $(6, 3)$.
$$r = \sqrt{(x_2 - x_1)^2 + (y_2 - y_1)^2} = \sqrt{(6 - 2)^2 + (3 - 3)^2} = \sqrt{4^2 + 0^2} = 4$$

10. What is an equation of a circle with its center at $(-3, 5)$ and a radius of 4?

(1) $(x - 3)^2 + (y + 5)^2 = 16$ (3) $(x - 3)^2 + (y + 5)^2 = 4$

*(2) $(x + 3)^2 + (y - 5)^2 = 16$ (4) $(x + 3)^2 + (y - 5)^2 = 4$

11. A circle is represented by the equation $x^2 + (y+3)^2 = 13$. What are the coordinates of the center of the circle and the length of the radius?

(1) $(0, 3)$ and 13 (2) $(0, 3)$ and $\sqrt{13}$ (3) $(0, -3)$ and 13 *(4) $(0, -3)$ and $\sqrt{13}$

12. What are the center and the radius of the circle whose equation is $(x-3)^2 + (y+3)^2 = 36$

*(1)   **center = $(3, -3)$; radius = 6**   (3)   center = $(3, -3)$; radius = 36

(2)   center = $(-3, 3)$; radius = 6   (4)   center = $(-3, 3)$; radius = 36

13. Which equation represents circle K shown in the graph below?

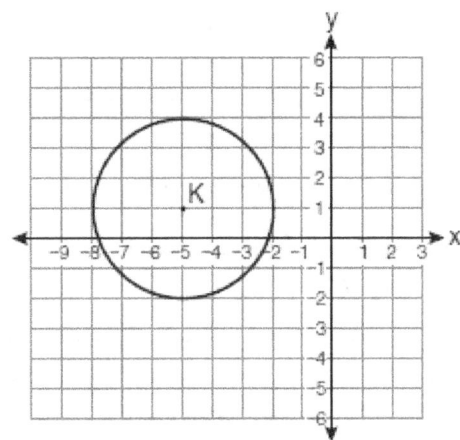

(1) $(x+5)^2 + (y-1)^2 = 3$ (3) $(x-5)^2 + (y+1)^2 = 3$

*(2) $(x+5)^2 + (y-1)^2 = 9$ (4) $(x-5)^2 + (y+1)^2 = 9$

Center = (-5, 1), radius = 3

---

14. Which equation represents the circle whose center is $(-2, 3)$ and whose radius is 5?

(1) $(x-2)^2 + (y+3)^2 = 5$ *(3) $(x+2)^2 + (y-3)^2 = 25$

(2) $(x+2)^2 + (y-3)^2 = 5$ (4) $(x-2)^2 + (y+3)^2 = 25$

15. Given the system of equations:

$$y = x^2 - 4x$$

$$x = 4$$

The number of points of intersection is

*(1)   1   (2)   2   (3)   3   (4)   0

Solve the system of equations by substitution.
$y = x^2 - 4x = (4)^2 - 4(4) = 16 - 16 = 0$
The point of intersection is (4, 0).

---

16. Given the equations: $y = x^2 - 6x + 10$

$$y + x = 4$$

What is the solution to the given system of equations?

(1) $(2, 3)$ (2) $(3, 2)$ (3) $(2, 2)$ and $(1, 3)$ *(4) $(2, 2)$ and $(3, 1)$

Solve the system of equations by substitution.

$y + x = 4$, $y = -x + 4$

$-x + 4 = x^2 - 6x + 10$

$x^2 - 5x + 6 = 0$

$(x - 2)(x - 3) = 0$

$x = 2$	$x = 3$
$y = -x + 4 = -2 + 4 = 2$	$y = -x + 4 = -3 + 4 = 1$
$(2, 2)$	$(3, 1)$

---

17. The equation of a circle is $(x - 2)^2 + (y + 4)^2 = 4$. Which diagram is the graph of the circle?

(1)   (3)

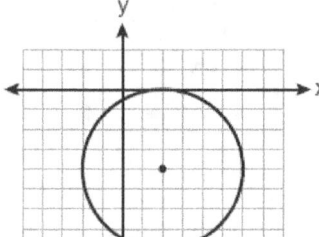 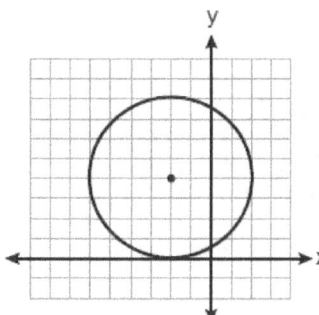

*(2)   (4)

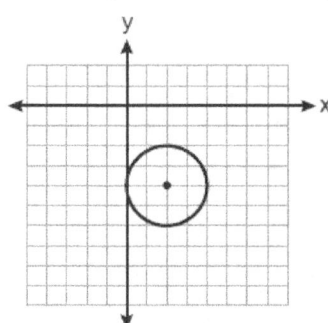 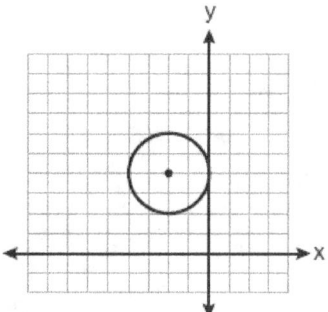

---

center = $(2, -4)$,   radius = 2

18. Which graph could be used to find the solution to the following system of equations?

$$y = -x + 2$$

$$y = x^2$$

(1)

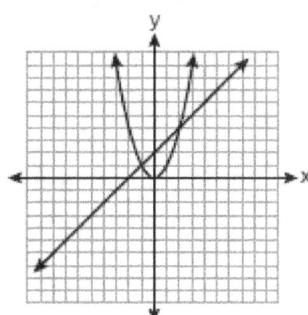

*(3)

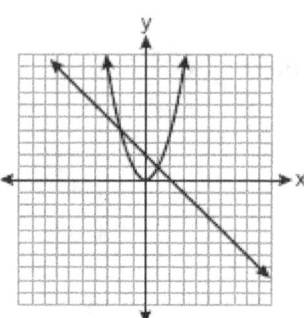

(2)

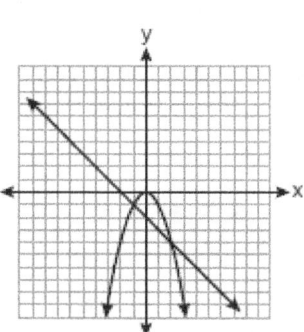

(4)

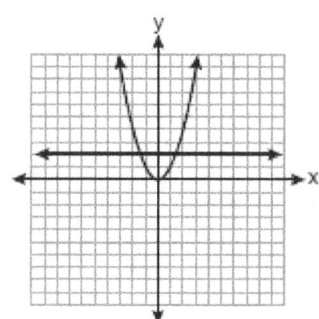

Show Work:

1. Using a compass and straightedge, construct the bisector of the angle shown below. [*Leave all construction marks.*]

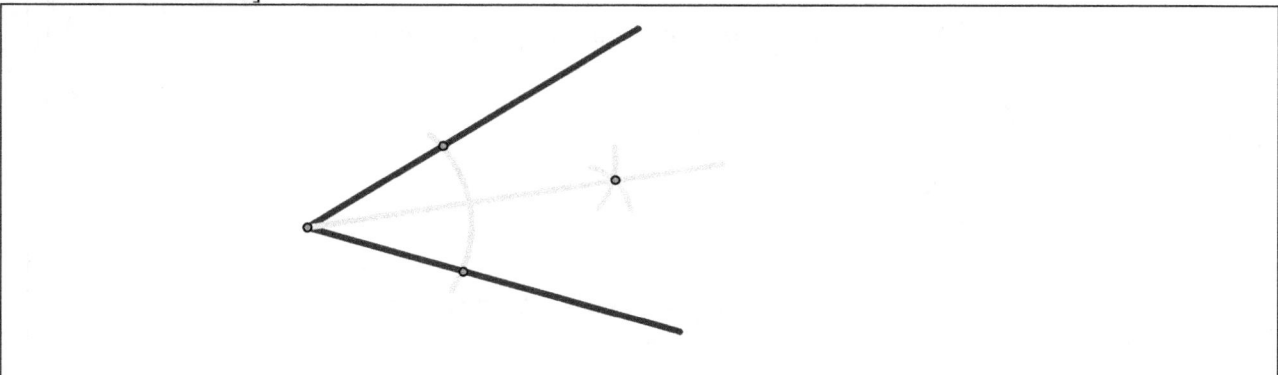

2. Using a compass and straightedge, construct a line that passes through point *P* and is perpendicular to line *m*. [Leave all construction marks.]

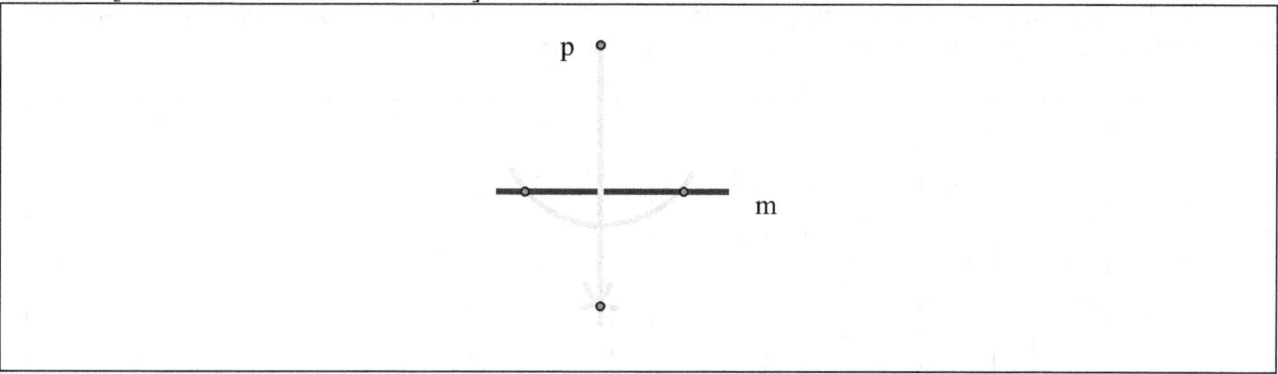

3. Using a compass and straightedge, and $\overline{AB}$ below, construct an equilateral triangle with all sides congruent to $\overline{AB}$. [Leave all construction marks.]

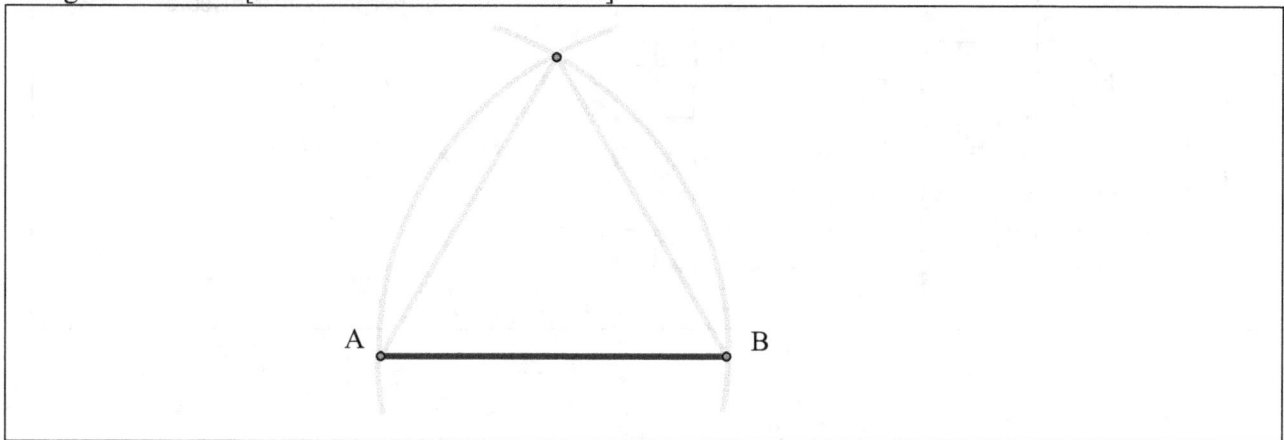

4. The length of $\overline{AB}$ is 3 inches. On the diagram below, sketch the points that are equidistant from A and B and sketch the points that are 2 inches from A. Label with an **X** all points that satisfy both conditions.

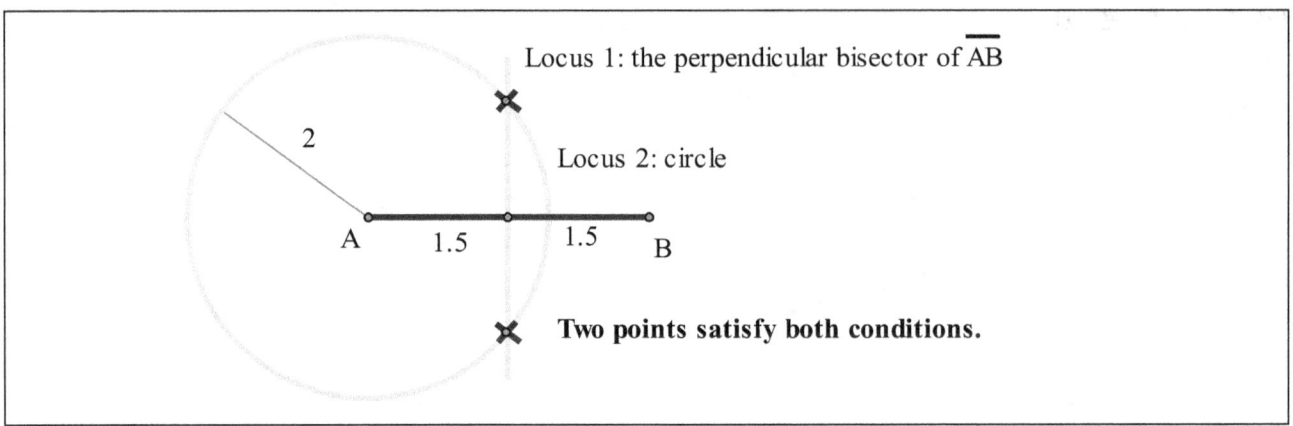

Locus 1: the perpendicular bisector of $\overline{AB}$

Locus 2: circle

Two points satisfy both conditions.

5. A city is planning to build a new park. The park must be equidistant from school A at $(3, 3)$ and school B at $(3, -5)$. The park also must be exactly 5 miles from the center of town, which is located at the origin on the coordinate graph. Each unit on the graph represents 1 mile. On the set of axes below, sketch the compound loci and label with an **X** all possible locations for the new park.

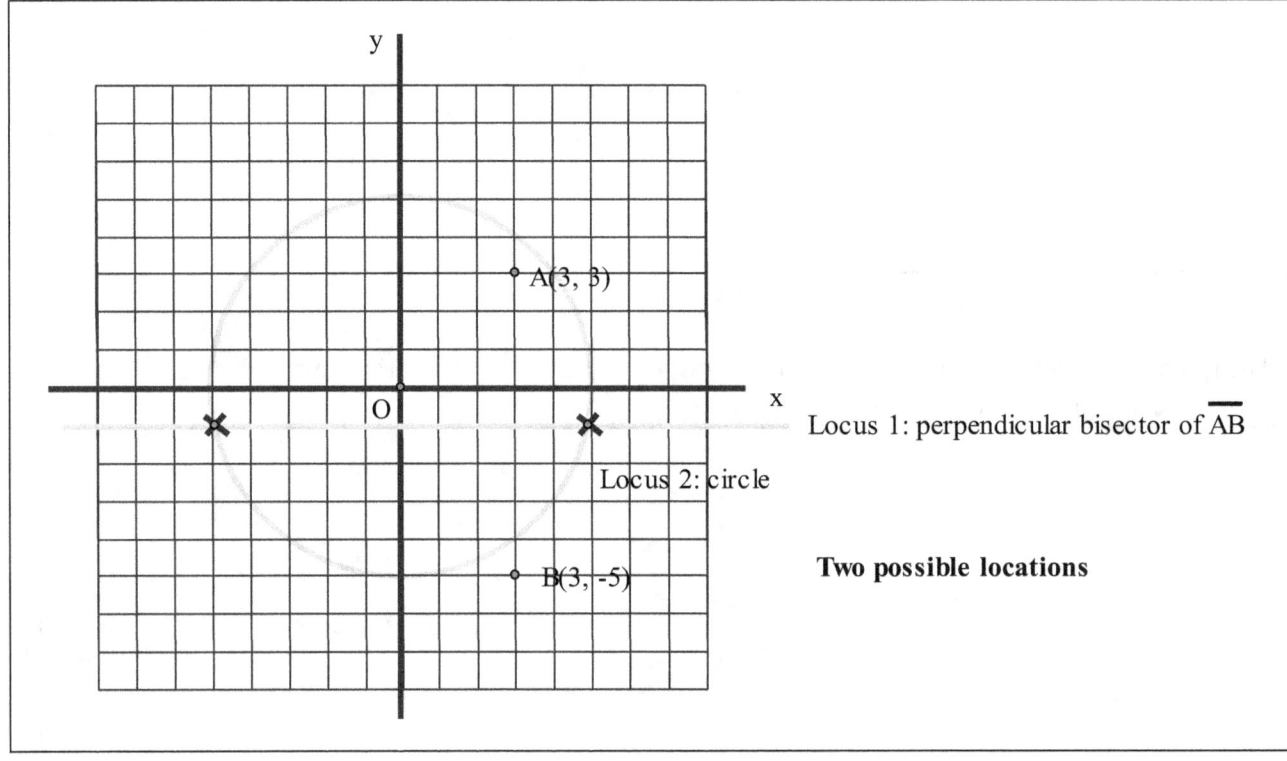

Locus 1: perpendicular bisector of $\overline{AB}$

Locus 2: circle

Two possible locations

6. On the grid below, graph the points that are equidistant from both the *x* and *y* axes and the points that are 5 units from the origin. Label with an **X** all points that satisfy both conditions.

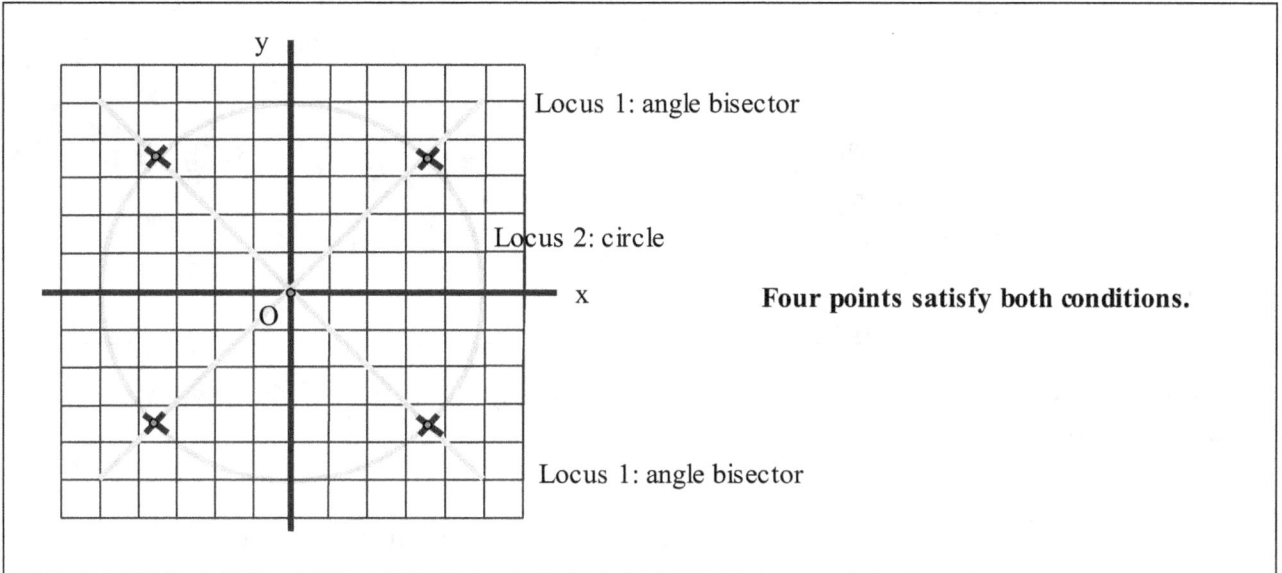

Locus 1: angle bisector

Locus 2: circle

Four points satisfy both conditions.

Locus 1: angle bisector

7. Write an equation of the perpendicular bisector of the line segment whose endpoints are $(-1, 1)$ and $(7, -5)$. [The use of the grid below is optional]

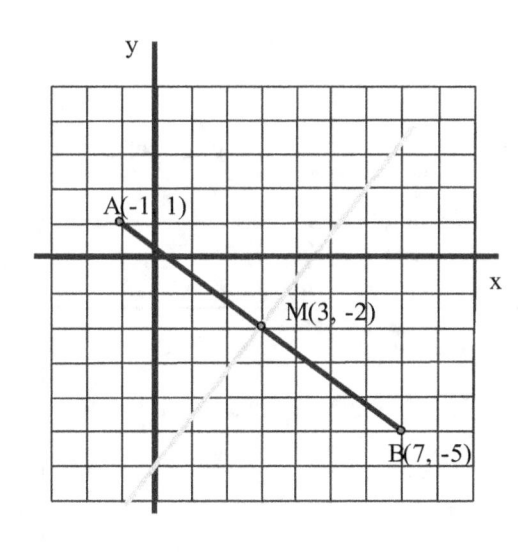

Find the midpoint of $\overline{AB}$:

$$M(x, y) = M\left(\frac{x_1 + x_2}{2}, \frac{y_1 + y_2}{2}\right) = M\left(\frac{-1 + 7}{2}, \frac{1 - 5}{2}\right)$$
$$= M(3, -2)$$

Find the slope of $\overline{AB}$:

$$m_1 = \frac{y_2 - y_1}{x_2 - x_1} = \frac{-5 - 1}{7 + 1} = \frac{-6}{8} = -\frac{3}{4}$$

the slope of the perpendicular line: $m_2 = -\dfrac{1}{m_1} = \dfrac{4}{3}$

the equation of the perpendicular bisector:

the slope is $\dfrac{4}{3}$ and passing through midpoint $(3, -2)$

the point-slope form: $y - y_1 = m(x - x_1)$

$$y + 2 = \frac{4}{3}(x - 3) \quad \text{(or slope-intercept form: } y = \frac{4}{3}x - 6 \text{)}$$

X. TRANSFORMATION

1. Triangle ABC has vertices $A(1, 3)$, $B(0, 1)$, and $C(4, 0)$. Under a translation, A', the image point of A, is located at $(4, 4)$. Under this same translation, point C' is located at

*(1) $(7, 1)$ (2) $(5, 3)$ (3) $(3, 2)$ (4) $(1, -1)$

A(1, 3) $T_{a, b}$ A'(1 + a, 3 + b) = A'(4, 4)

                      1 + a = 4 ,   3 + b = 4
                      a = 3 ,   b = 1

C(4, 0) $T_{3, 1}$ C'(4 + 3, 0 + 1) = C'(7, 1)

---

2. A polygon is transformed according to the rule: $(x, y) \rightarrow (x + 2, y)$. Every point of the polygon moves two units in which direction?

(1) up   (2) down   (3) left   ***(4)** **right**

3. In the diagram below, under which transformation will $\triangle A'B'C'$ be the image of $\triangle ABC$?

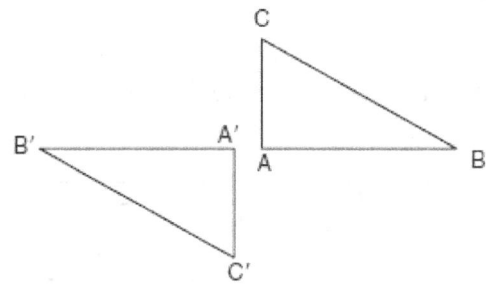

*(1) **rotation**   (2) dilation   (3) translation   (4) glide reflection

4. Point A is located at $(4, -7)$. The point is reflected in the x-axis. Its image is located at

(1) $(-4, 7)$ (2) $(-4, -7)$ ***(3)** $(4, 7)$ (4) $(7, -4)$

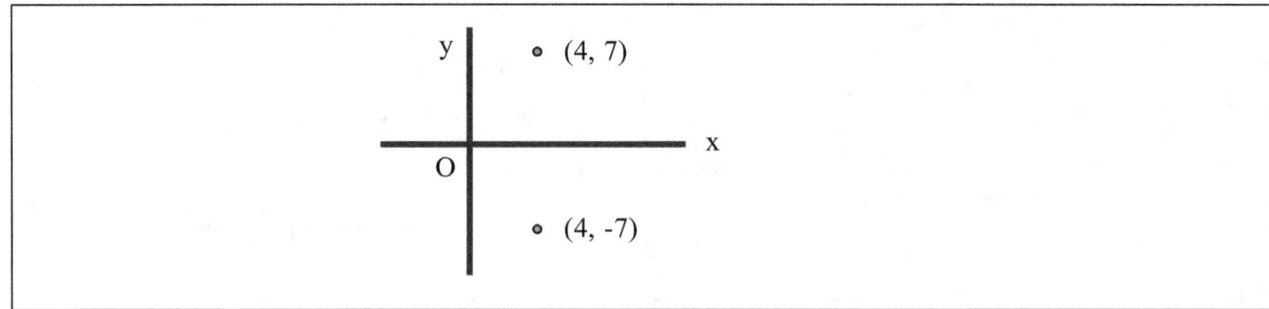

5. Which transformation produces a figure similar but not congruent to the original figure?

(1) $T_{1,3}$ ***(2)** $D_{\frac{1}{2}}$ (3) $R_{90°}$ (4) r_{y-x}

6. On the set of axes below, Geoff drew rectangle *ABCD*. He will transform the rectangle by using the translation $(x,y) \rightarrow (x+2, y+1)$ and then will reflect the translated rectangle over the *x*-axis.

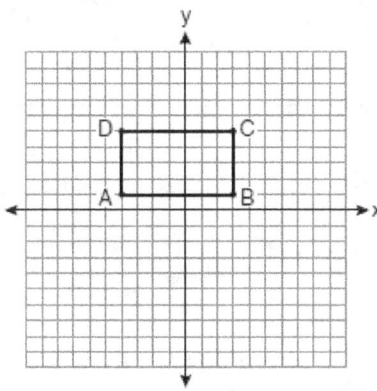

What will be the area of the rectangle after these transformations?

*(1) **exactly 28 square units**   (3) greater than 28 square units

(2) less than 28 square units   (4) It cannot be determined from the information given.

> Only dilation enlarges or reduces the size of the image, which is similar to the original.
> The image of other transformations is congruent to the original.

7. Which expression best describes the transformation shown in the diagram below?

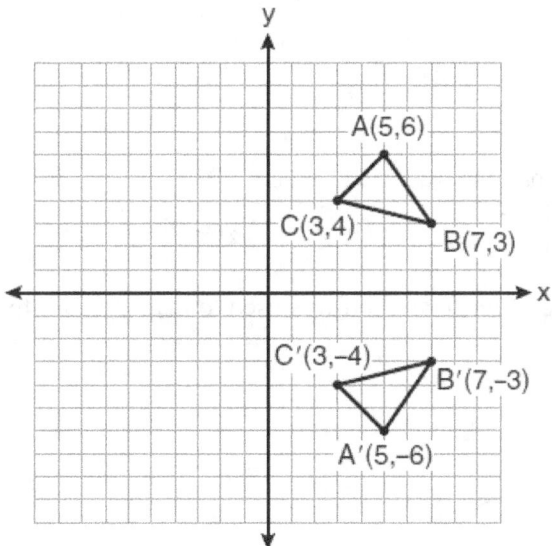

(1) same orientation; reflection   (3) same orientation; translation

*(2) **opposite orientation; reflection**   (4) opposite orientation; translation

8. The endpoints of $\overline{AB}$ are $A(3,2)$ and $B(7,1)$. If $\overline{A''B''}$ is the result of the transformation of $\overline{AB}$ under $D_2 \circ T_{-4,3}$ what are the coordinates of A'' and B''?

*(1) $A''(-2,10)$ **and** $B''(6,8)$ (3) $A''(2,7)$ and $B''(10,5)$

(2) $A''(-1,5)$ and $B''(3,4)$ (4) $A''(14,-2)$ and $B''(22,-4)$

$T_{-4,3}$ followed by D_2

$A(3,2)$ $\underline{\quad T_{-4,3} \quad}$ $A'(3-4, 2+3) = A'(-1, 5)$ $\underline{\quad D_2 \quad}$ $A''(2\bullet(-1), 2\bullet5) = A''(-2, 10)$

$B(7,1)$ $\underline{\quad T_{-4,3} \quad}$ $B'(7-4, 1+3) = B'(3, 4)$ $\underline{\quad D_2 \quad}$ $B''(2\bullet3, 2\bullet4) = B''(6, 8)$

---

9. After a composition of transformations, the coordinates $A(4,2)$, $B(4,6)$, and $C(2,6)$ become $A''(-2,-1)$, $B''(-2,-3)$, and $C''(-1,-3)$, as shown on the set of axes below.

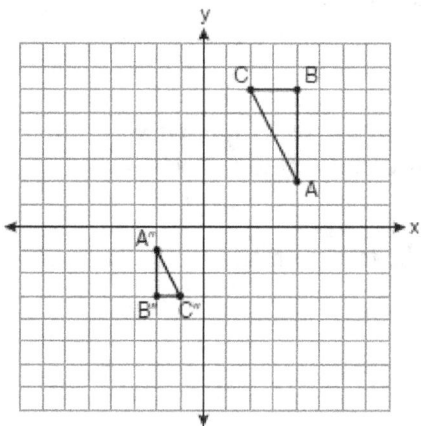

Which composition of transformations was used?

(1) $R_{180°} \circ D_2$ (2) $R_{90°} \circ D_2$ *(3) $D_{\frac{1}{2}} \circ R_{180°}$ (4) $D_{\frac{1}{2}} \circ R_{90°}$

10. In the diagram below, which transformation was used to map $\triangle ABC$ to $\triangle A'B'C'$?

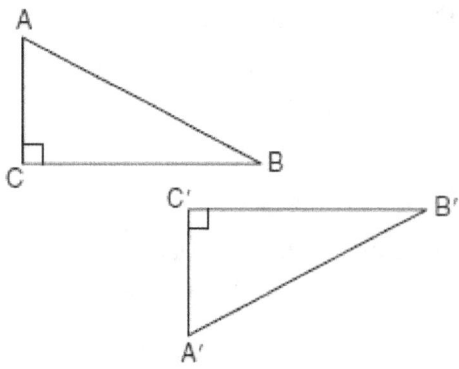

(1) dilation   (2) rotation   (3) reflection   *(4) **glide reflection**

11. What is the image of point $A(4, 2)$ after the composition of transformations defined by $R_{90°} \circ r_{y=x}$?

*(1) $(-4, 2)$ **(2)** $(4, -2)$ **(3)** $(-4, -2)$ **(4)** $(2, -4)$

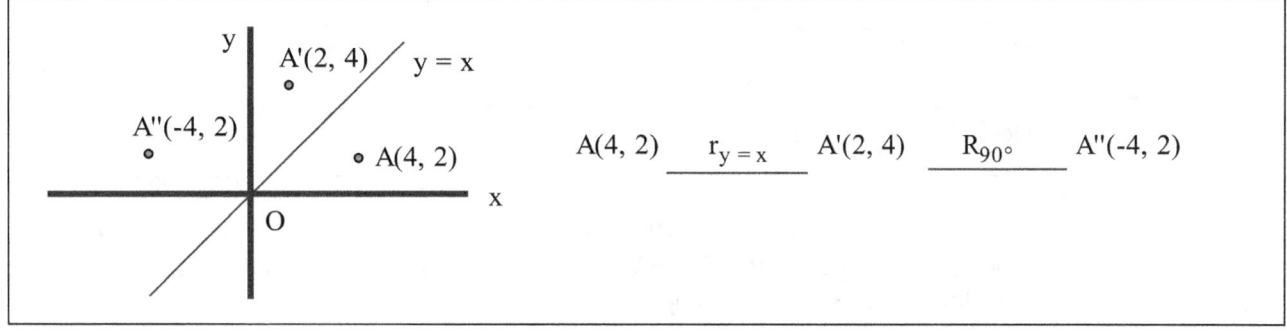

12. Which transformation is *not* always an isometry?

(1)   rotation   ***(2)   dilation**   (3)   reflection   (4)   translation

Show Work:

1. Triangle *DEG* has the coordinates $D(1, 1)$, $E(5, 1)$, and $G(5, 4)$. Triangle *DEG* is rotated 90° about the origin to form $\triangle D'E'G'$. On the grid below, graph and label $\triangle DEG$ and $\triangle D'E'G'$. State the coordinates of the vertices D', E', and G'. Justify that this transformation preserves distance.

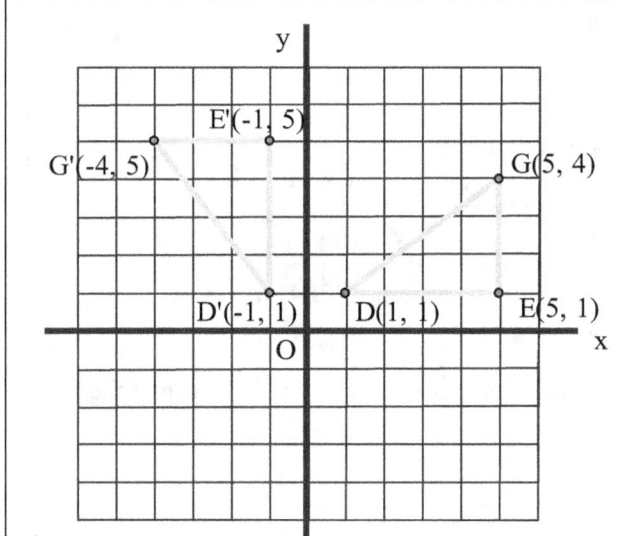

DE = D'E' = 4

EG = E'G' = 3

GD = G'D' = 5

Rotation preserves distance.

2. In $\triangle KLM$, $m\angle K = 36$ and $KM = 5$. The transformation D_2 is performed on $\triangle KLM$ to form $\triangle K'L'M'$. Find $m\angle K'$. Justify your answer. Find the length of $\overline{K'M'}$. Justify your answer.

$m\angle K' = m\angle K = 36$; $K'M' = 2 \bullet KM = 2 \bullet 5 = 10$

Under a dilation $\triangle K'L'M'$ is similar to $\triangle KLM$. Their corresponding angles are congruent and their corresponding sides are in proportion, here it is 2 to 1.

---

3. The coordinates of the vertices of parallelogram $ABCD$ are $A(-2, 2)$, $B(3, 5)$, $C(4, 2)$, and $D(-1, -1)$. State the coordinates of the vertices of parallelogram $A''B''C''D''$ that result from the transformation $r_{y-axis} \circ T_{2,-3}$. [The use of the set of axes below is optional.]

$A(-2, 2)$ $\underline{T_{2,\,-3}}$ $A'(-2 + 2, 2 - 3) = A'(0, -1)$ $\underline{r_{y-axis}}$ $A''(0, -1)$

$B(3, 5)$ $\underline{T_{2,\,-3}}$ $B'(3 + 2, 5 - 3) = B'(5, 2)$ $\underline{r_{y-axis}}$ $B''(-5, 2)$

$C(4, 2)$ $\underline{T_{2,\,-3}}$ $C'(4 + 2, 2 - 3) = C'(6, -1)$ $\underline{r_{y-axis}}$ $C''(-6, -1)$

$D(-1, -1)$ $\underline{T_{2,\,-3}}$ $D'(-1 + 2, -1 - 3) = D'(1, -4)$ $\underline{r_{y-axis}}$ $D''(-1, -4)$

---

4. On the set of axes below, solve the following system of equations graphically for all values of x and y.

$$y = (x - 2)^2 + 4$$

$$4x + 2y = 14$$

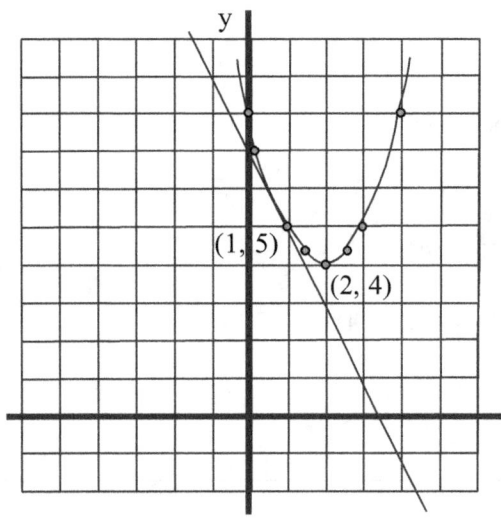

A function under a translation:

$$y = f(x) \quad T_{a,\,b} \quad y = f(x - a) + b$$

$$y = x^2 \quad T_{2,\,4} \quad y = (x - 2)^2 + 4$$

The vertex $(0, 0)$ of $y = x^2$ becomes $(2, 4)$ of $y = (x - 2)^2 + 4$ under the translation.

To graph the parabola, properly choose 5 points around the vertex $(2, 4)$

x	0	1	2	3	4
y	8	5	4	5	8

To graph $4x + 2y = 14$, write it in Slope-Intercept form: $2y = -4x + 14$, $y = -2x + 7$

The solution of the system of equations is the point of intersection $(1, 5)$ which satisfies both equations.

XI. SOLID GEOMETRY

1. Lines k_1 and k_2 intersect at point E. Line m is perpendicular to lines k_1 and k_2 at point E.

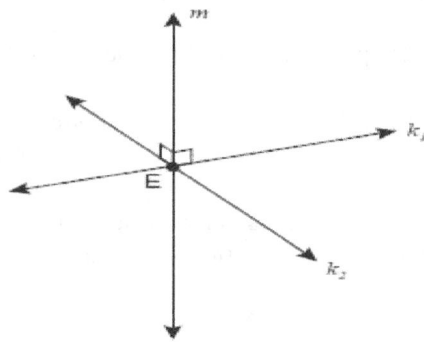

Which statement is always true?
(1) Lines k_1 and k_2 are perpendicular.
(2) Line m is parallel to the plane determined by lines k_1 and k_2.
*(3) Line m is perpendicular to the plane determined by lines k_1 and k_2.
(4) Line m is coplanar with lines k_1 and k_2.

2. Point P is on line m. What is the total number of planes that are perpendicular to line m and pass through point P?
*(1)   1   (2)   2   (3)   0   (4) infinite

3. If two different lines are perpendicular to the same plane, they are
(1) collinear   ***(2) coplanar**   (3) congruent   (4) consecutive

4. In plane $\mathcal{P}$, lines m and n intersect at point A. If line k is perpendicular to line m and line n at point A, then line k is
(1)   contained in plane $\mathcal{P}$ ***(3) perpendicular to plane $\mathcal{P}$**
(2)   parallel to plane $\mathcal{P}$ (4) skew to plane $\mathcal{P}$

5. In the diagram below, line k is perpendicular to plane $\mathcal{P}$ at point T.

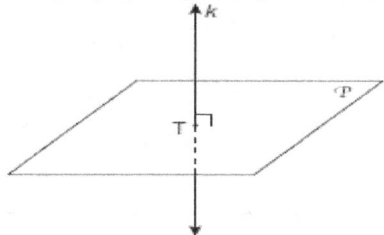

Which statement is true?
(1) Any point in plane $\mathcal{P}$ also will be on line k.
(2) Only one line in plane $\mathcal{P}$ will intersect line k.
(3) All planes that intersect plane $\mathcal{P}$ will pass through T.
*(4) Any plane containing line k is perpendicular to plane $\mathcal{P}$.

6. Line *k* is drawn so that it is perpendicular to two distinct planes, *P* and *R*. What must be true about planes *P* and *R*?

(1) Planes *P* and *R* are skew.   ***(2) Planes *P* and *R* are parallel.**

(3) Planes *P* and *R* are perpendicular.   (4) Plane *P* intersects plane *R* but is not perpendicular to plane *R*.

7. In three-dimensional space, two planes are parallel and a third plane intersects both of the parallel planes. The intersection of the planes is a

(1) plane   (2) point   ***(3) pair of parallel lines**   (4) pair of intersecting lines

8. Through a given point, *P*, on a plane, how many lines can be drawn that are perpendicular to that plane?

*(1)   1   (2)   2   (3)   more than 2   (4)   none

XIII. GEOMETRIC MEASUREMENTS

1. The figure in the diagram below is a triangular prism.

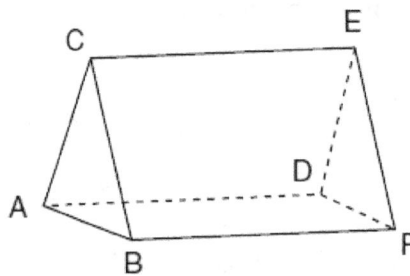

Which statement must be true?

(1) $\overline{DE} \cong \overline{AB}$ (2) $\overline{AD} \cong \overline{BC}$ ***(3) $\overline{AD} \parallel \overline{CE}$**   (4) $\overline{DE} \parallel \overline{BC}$

2. A right circular cylinder has a volume of 1,000 cubic inches and a height of 8 inches. What is the radius of the cylinder to the *nearest tenth of an inch*?

*(1)   6.3   (2)   11.2   (3)   19.8   (4)   39.8

$$V = \pi r^2 h$$
$$1000 = \pi r^2 \bullet 8$$
$$r = \sqrt{\frac{1000}{8\pi}} \approx 6.3$$

3. The lateral faces of a regular pyramid are composed of

(1) squares   (2) rectangles   (3) congruent right triangles   ***(4)   congruent isosceles triangles**

4. Which expression represents the volume, in cubic centimeters, of the cylinder represented in the diagram below?

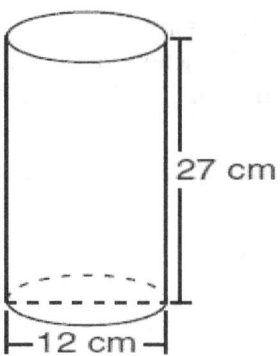

27 cm

12 cm

(1) 162π (2) 324π *(3) 972π (4) $3,888\pi$

Diameter = 12 , Radius = 6 , $V = Bh = \pi r^2 h = \pi \cdot 6^2 \cdot 27 = 972\pi$

---

5. In the diagram below, a right circular cone has a diameter of 8 inches and a height of 12 inches.

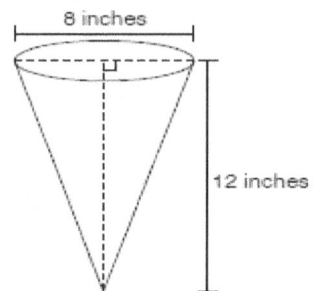

8 inches

12 inches

What is the volume of the cone to the *nearest cubic inch*?
*(1)   **201**   (2)   481   (3)   603   (4)   804

---

d = 8 , r = 4 , h = 12

$V = \dfrac{1}{3}\pi r^2 h = \dfrac{1}{3}\pi \bullet 4^2 \bullet 12 \approx 201$

Show Work:

1. Tim has a rectangular prism with a length of 10 centimeters, a width of 2 centimeters, and an unknown height. He needs to build another rectangular prism with a length of 5 centimeters and the same height as the original prism. The volume of the two prisms will be the same. Find the width, in centimeters, of the new prism.

$$V = l \bullet w \bullet h$$
(old)　$10 \bullet 2 \bullet h = 5 \bullet w \bullet h$　(new)
$$20 = 5w$$
$$\mathbf{w = 4}$$

2. The volume of a cylinder is 12,566.4 cm^3. The height of the cylinder is 8 cm. Find the radius of the cylinder to the *nearest tenth of a centimeter*.

$$V = \pi r^2 h$$
$$12566.4 = \pi r^2 \bullet 8$$
$$r = \sqrt{\frac{12566.4}{8\pi}} \approx \mathbf{22.4}$$

3. A regular pyramid with a square base is shown in the diagram below.

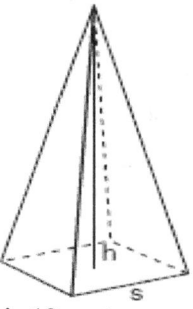

A side, *s*, of the base of the pyramid is 12 meters, and the height, *h*, is 42 meters. What is the volume of the pyramid in cubic meters?

$$V = \frac{1}{3}Bh = \frac{1}{3}s^2h$$
$$= \frac{1}{3} \bullet 12^2 \bullet 42$$
$$= \mathbf{2016}$$